beneficial
CONSULTING

To my wife Tatiana for your inspiration, encouragement
and support.
Ты наполняешь мою душу радостью.

Al Shalloway, Jérôme Tassel, and Ziaan Hattingh
Thanks for your suggestions, mentoring, and encouragement!

Agile Lineout

A framework for change teams

Jonathan Ward
Beneficial Consulting Ltd.
March 2023

Contents

Introduction

This book is about the implementation of strategic change. Its Lean-Agile modern management practices are designed to help organisations close the gap between strategic intent and reality.

When coaching agile teams engaged in software engineering, I have often been approached by people with a marketing, legal or HR background and asked if they can do agile and, if so, how. I wrote this book as a response to these requests.

This booklet describes a framework which I have called Agile Lineout. Its contents describe agile and lean practices with modern management's mindset and behavioural theories.

I wanted to return to the original academic materials. But in exploring these resources, I discovered that in some instances, popular agile wisdom has adapted the original theories or science. Unfortunately, in some cases, this adaptation is not for the better!

I hope that by outlining the theory with the practice, I will enable teams to self-improve and learn based on their understanding of the concepts and techniques and their observations when applying these to their unique project context.

Many organisations suffer from what has become known as the Strategy Execution gap. This gap is where an organisation encounters difficulty converting strategic plans into reality. I have designed Agile Lineout to facilitate the execution of strategic plans using a quarterly big-room planning approach with rolling 30-60-90 day planning.

This booklet has a broad audience. However, it primarily aims to assist individuals and teams in visualising, clarifying and implementing Strategic Change activities. However, it may also be applied in large and small businesses with strategic activities to transform, adapt, merge, or create new enterprise functions, HR activities and marketing undertakings.

People often struggle to adapt agile software engineering practices to fit non-tech or strategic change activities. This guide removes the need to translate software techniques by outlining how to use Lean-agile modern management principles with Strategic Change activities. It may also be used with activities that combine software development with strategic change. Consequently, Agile Lineout, although designed for strategic change, can also be used for pure software engineering activities.

People will get various potential benefits from reading this book:

- **Business stakeholders** will learn more effective ways of delivering strategy, engaging with agile teams, and enabling faster business outcomes by delivering key results.
- **Solution owners** will understand how to break strategic change activities into increments, increasing speed and the value delivered.
- **Team members** will understand why their involvement in strategic change is essential. In addition, why they should focus on their effectiveness and how they should adapt their plans based on feedback and interpretation of progression metrics.
- **Team coaches** will know how to support, coach and improve team performance using Agile Lineout— benefiting from its three lines of guidance; solution, development and team behaviour. This booklet also contains specific advice in the coaching chapter for team coaches introducing Agile Lineout.
- **Agile Programme Managers** will understand that scaling Agile Lineout requires no additional learning beyond operating as several teams. Involving a greater understanding of the complexity and progression of the solution they are creating and the team dynamics used to create it.

What is Agile?

I am often asked what is agile. It is quite a difficult question to answer. I tend to use the following headline statement; Agile is a different and constantly evolving modern management way of working. It uses the organisational construct of teams, harnessing collective intelligence and empowering people with devolved authority and decision-making. I then find that I need to explain the nuances of my statement and illustrate lean-agile practices using examples.

The primary difference between traditional project management and Agile is the psyche or mindset. The conventional project management approach focuses on delivery within time and cost and is based on the following system.

Process →   Solution →   People

The first concern of a project manager is to create a detailed plan which defines the work breakdown structure and delivery process. Then, the solution will be explored early in the project lifecycle, and the requirements will be determined. Next comes the estimation process, which layers the constraints of skillsets, effort and team capacity onto the plan's tasks. In contrast to the agile approach, estimation is usually the responsibility of someone outside of the development team. The result is a detailed project timeline which, forever after that, is locked down and used for variance analysis and reporting.

The documentation used in a traditional project detaches the project value contained in a business case from the functionality and systems requirements defined in later documents. So if the planned timetable or budget is forecast to be exceeded, then it is commonplace for the scope of the solution to be adjusted. Consequently, functions that may provide significant value to a customer are often not delivered.

The Agile mindset, in contrast, places emphasis on value creation. From the outset, the agile approach focuses on the value of the outcome for the customer. It is based on the following system.

Value →   People →   Process

The first concern of an agile team is to break their solution into valuable components ordered by value in what is known as a backlog. Agile teams deliver solutions incrementally. If implemented, each increment must be viable. Being viable means creating value for the customer and not exposing the organisation to operational risk or compliance issues.

The team estimates each increment based on its functionality to create the solution roadmap or summary delivery plan. Next, they place the increments which deliver the highest value at the top of the list. The team then pulls work from the backlog when they can start working on the increment. In this way, agile uses a product breakdown approach rather than a work breakdown used in traditional project management.

Furthermore, an agile team is self-organising, a team of equals who decide how to create the required outcome. There is typically no manager within a team but collective responsibility for decision-making. Although a team will have informal leaders, they assign roles among themselves. An agile team may have a coach, but this position does not have management authority.

An agile team works in short-term cycles of one or two weeks, sometimes a month, but rarely longer. When work has been selected for development, the highest-value items receive the first attention by creating a detailed plan for the next cycle. Then, by breaking down the next highest value item into tasks, estimating those tasks, and aligning the workload to capacity, the capabilities and work capacity of the people are given higher importance than the development process.

The contrast between traditional project management with its detailed process plan and agile with its detailed product breakdown and high-level solution road map is quite stark.

Another contrast is the focus in agile on the teamwork needed to develop a valuable solution, product, or service. Initially, a team is assembled around a vision or objective and organises itself to deliver incrementally. Agile is designed to engage a team's collective intelligence and energy rather than that of a single individual or project manager. The work could be strategic change, a marketing strategy, new technology, or software. Actually, it can be anything that adds value to the organisation to which the team belongs.

Why not use Scrum?

The most popular agile framework in use by far is Scrum, so why not use it for strategic change activities too? But unfortunately, it is because of how Scrum operates, designed for software engineering, which makes it inappropriate.

For its efficacy, Scrum relies upon a team's ability to take a task and get it to a definition of done within a single sprint of maybe two- or three-week duration. Completing work items within a sprint means that tasks can be reviewed and the work plan adapted in the next sprint so that the team achieves their goal. Unfortunately, many tasks performed in strategic change activities cannot be assessed as being complete in such a short window.

Take a cultural change activity using Scrum as an example. The effect of some tasks undertaken by the team may not be apparent for two or three months, sometimes even longer. So how does that Scrum team know their activity is complete in a single sprint? Breaking tasks into segments may be one way of doing strategic change using Scrum, but splitting the work from being able to observe the result would be necessary. In this case, how would a team get a meaningful picture of their completed work? How does this give the team the transparency they need?

On the other hand, if a cultural change team uses Kanban as an alternative, The team can move the task ticket into a monitoring column on their Kanban board and have transparency. The ticket in the monitoring column reminds the team that the effect of a task is being or has not yet been assessed. A task could stay in that monitoring column for weeks or months, but being there always reminds the team that they must do something. Kanban creates the transparency a cultural change team needs to control its activity. Not being able to get to the "done" state in a short time window is my primary reason for recommending not to use Scrum.

Scrum also relies upon software engineering disciplines from extreme programming (XP), Lean and other sources to make it workable. For example, how to estimate, define stories, and create acceptance criteria. The Scrum Guide 2020 states, *"the Scrum framework is purposefully incomplete, only defining the parts required to implement Scrum theory"*. Being purposefully incomplete, Scrum challenges many teams to decide what additions they need for a complete agile framework.

The techniques used by software teams require significant adaptation if strategic change teams are to use them. In Agile Lineout, I have therefore sought to provide a more complete guide yet still left it open to teams to make decisions based on their context, business objectives, and delivery needs. As a result, Agile Lineout supports the whole change delivery system.

New to agile teams of all types, not understanding the concepts of self-management and collective responsibility, often wait for instructions and leadership. So Agile Lineout steers those responsible for initiating activities, building teams and pointing them towards defining their strategic objective or solution, the development process they will use to create the outcome, and the steps to working as a team. The level of support provided enables focus on the business objective, rapid team start-up and facilitates faster delivery based on the continued assessment of the context of their activity.

The Lineout Paradigm

The term Scrum has its origins in the game of rugby football. In their paper entitled "The New New Product Development Game", Hirotaka Takeuchi and Ikujiro Nonaka explained the connection: *"Under the rugby approach, the product development process emerges from the constant interaction of a hand-picked, multi-disciplinary team whose members work together from the start to finish. Rather than moving in defined, highly structured stages, the process is born out of the team members' interplay"*. A scrum is one of rugby's set pieces; a lineout is another. In rugby football, the team interplay is the same once Scrum or Lineout restarts the game.

In a rugby lineout, both teams assemble side-by-side. Although both teams try to win the ball, the attacking side behaves differently from the defending side. Both teams have practised plans and strategies to bring into play when the ball is thrown between the two lines to restart the game. Once the rugby lineout is complete, each team acts in concert to defend, attack, and win the game. An agile team interacts similarly using Agile Lineout as their starting point. Agile Lineout provides recommended steps and techniques, but each team must decide the game they wish to play in their context.

Many agile frameworks teach their approach by rote, and the learning process is based on practice and repetition rather than the underlying theory. For example, some SAFe examination questions insist on using the training deck's actual words to gain certification! Learning by rote often limits understanding, particularly where underlying management science or theory is being applied. In Agile Lineout, I have made explicit the management theory so individuals and teams can reflect, learn and improve having understood the intent.

In the New Economics for Industry, Dr W. Edwards Deming wrote, *"experience by itself teaches nothing... Without theory, experience has no meaning. Without theory, one has no questions to ask. Hence, without theory, there is no learning."* Deming's assertion is borne out by my experience of Scrum Masters unwilling to experiment with alternative patterns to increase the performance of their teams because it wasn't taught in their certification classes. I believe that if we teach theory combined with the practices of each framework, including Kanban and Scrum, individuals and teams will have the knowledge and confidence to enable them to improve continuously.

System of profound knowledge

Agile Lineout uses Dr Deming's system of profound knowledge. Profound knowledge is a management philosophy grounded in systems theory. The theory asserts that every system has four elements. Namely:

- **Understanding the system:** The system is how people interact using tools and techniques to deliver an outcome. In a system, action in one part of the process affects the other parts.
- **Knowledge of Variation**: Variation in this context establishes why something turned out differently from what was planned. Knowing that variation is inevitable and having the determination to discover why differences have occurred enables learning and allows for choosing a different path in the future.
- Awareness that there is **no knowledge without theory**: Creating a plan or definition of an expected outcome establishes a hypothesis, a prediction. Learning may result when a variation of this hypothesis occurs, and it is investigated. Rational prediction requires theory and builds upon knowledge through systematic observation and revision. Furthermore, the use of theory encourages adaptation when the context of the activity changes.
- **Understand human psychology**: this element describes the interaction between work systems and people, involving questions such as: How do people learn? What motivates people? How do people relate to each other in teams?

Using Agile Lineout, a team understands their strategic objective and then creates a plan. The Plan is their theory of how their activity may unfold. Then, they undertake activities and inevitably encounter variation – things rarely work as planned! In retrospection, during the Enhancement Workshop, the team deeply explores why things didn't turn out as planned. The cause of this variation may have political, technical or teamwork sources. In this exploration, they learn and then try things differently next time to increase predictability. I have designed Agile Lineout using these four profound knowledge components, placing them in each lineout step to encourage teams to learn and improve performance. Improving performance also involves assessing each element of the team's delivery system.

Systems Theory

Systems theory seeks to explain how elements in a complex system interact. In most organisations, strategic change only results from harnessing a complex system. The complexity lies in the new strategy and the means of attainment or delivery system. For example, in every strategic change or development activity, whether waterfall or agile, software or non-software, three components interact in the delivery system (see Figure 1) and are interdependent. These are:

Figure 1 The Delivery System

- **the solution itself** - the value or desired strategic outcome, the product or service,
- **the teamwork and behaviours** used to collaborate and produce the solution efficiently,
- **the tools and techniques,** the development process or the means of production used to create the customer value or solution.

As shown in Figure 1, the three interacting strategic change components are the solution or desired outcome, the means to develop that solution, i.e., the techniques and tools, and the team, which includes the roles, responsibilities, and relationships between the people necessary to achieve their goal.

To explain, when the solution or desired outcome is considered, the skills needed in team members will become apparent. The needs of the solution and constraints around the people involved will dictate the tools required. For example, if team members are spread across the globe, they will have different communication and collaboration needs than a collocated team.

Systems theory uses positive and negative feedback loops to encourage the system to adjust to its environment. In Agile Lineout, we have formalised the feedback loops. The Iteration Review, the

14

focus on Team Learning, the two sets of Inspection activities during each iteration and the Contextual Assessment at the beginning of each Iteration are all feedback mechanisms for the team.

Regardless of the application, the Agile Lineout system is designed to convert inputs into valuable strategic outputs. The fundamental concept of systems theory is that the whole is greater than the sum of its elements and that a system exists in an environment. The system is contextual.

Consider your last project and list the elements or parts involved. If another person looks at your list, they cannot tell whether the project was successful. Only once the context is explained may a third party be able to begin to evaluate the organisation of the team and the potential outcome of the activity. Internal and external factors influence the solution or successful delivery of the desired result, the development process, teamwork, and the complete delivery system. This fact explains why the Contextual Assessment is an essential and prominent element of Agile Lineout.

Every development activity has a strategic goal, product, solution, or outcome defined in some form of a business case or investment proposal. It is why the organisation has assembled a group of people or provided additional funding for an existing team. Agile Lineout offers a set of techniques to clarify the strategic goal or desired outcome, then break it down into actional work items, and monitor and review the key results.

Occasionally a team is brought together, and their first task is to define the desired result of their activity; some call this exploratory agile. In other instances, an agile team has a fixed objective, causing the team to be inventive about achieving the outcome rather than altering the goal itself. In both cases, the team needs to be clear about the outcome and receive regular and frequent feedback from stakeholders that they are on the right path. Agile Lineout, therefore, contains strategies and foundations to define, create and refine exploratory solutions with ill-defined outcomes.

Every strategic change activity has a process by which it creates its strategic outcome. It may not be a software engineering process; it could be designing and delivering the activities in a marketing strategy or restructuring a business. Each activity has a means of production, a value stream of tools, techniques, and methods that are a part of the system to develop the desired business outcome. In some situations, this value stream may be more apparent than in others.

Each team needs to be clear about how they will produce their solution, what skills they need, and the techniques and tools they will employ. They will also need to define their approach to quality and how they will ensure the variations to their intent are resolved. Agile Lineout contains strategies and foundations that are kept in focus so that a team may become proficient in delivering strategic change solutions.

Lastly, the team must work together as effectively as possible; they are part of the delivery system. A team will define roles and responsibilities and plan the capabilities they must develop or learn as they work together. Teamwork rarely happens by accident, yet working as a team is essential to successful project work and solution development. Agile Lineout contains strategies and foundational principles for establishing teams, building teams from working groups, gauging and improving team behaviour and encouraging performance.

As the activity progresses and the team better understands the scope of the work needed, a team may decide to alter its delivery process or how they wish to work together. Therefore, I have inbuilt the Enhancement Workshop and Contextual Assessment in Agile Lineout as steps when performance can be evaluated, and strategic adjustments to the solution, development process, or team working can be made.

I have used capital letters throughout this book to indicate an Agile Lineout, artefact, strategy, or foundational focus.

Agile Lineout

As previously stated, a delivery system has three elements: The Solution, The Development techniques and tools and the Team. In Agile Lineout, each element is described as a line (see Figure 2), and each line interacts with the others. It's a system!

- **The Solution Line** identifies strategies to create the strategic outcome and the foundations. It represents a continual team focus on creating business value.
- **The Development Line** outlines the tools, techniques and strategies to create an efficient delivery system and represents a continual team focus on effective development.
- **The Team Behaviour Line** enables both solution creation and development efficiency. In addition, the Team Behaviour Line has strategies for team building, enhancing teamwork and optimising delivery performance. These are the behavioural foundations that lay the bedrock for outstanding team performance.

In rugby, a team's lineout strategies and tactics vary by position on the field. Therefore, rugby players adjust their behaviour based

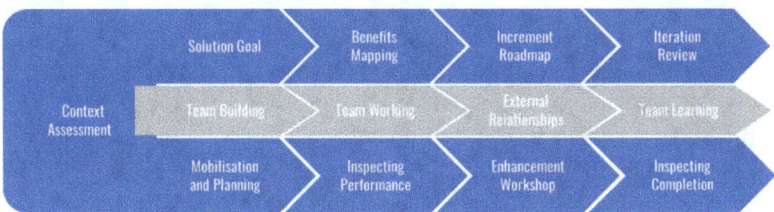

Figure 2 Agile Lineout

on the context of the game. Using Agile Lineout, an agile team mirrors this contextual adjustment in rugby using plan-do-study-adjust cycles and feedback loops. The agile team adjusts what it is doing and how it is doing based on its learning, internal observations, outcomes, and external feedback.

At the beginning of each Agile Lineout, the team assesses their context and decides on the next iteration's primary concerns.

In the Context Assessment, a team also evaluates whether the team behaviour (e.g., roles and responsibilities) and the development process (e.g., tools and ways of working) remain aligned with the solution goals or objectives when the latest stakeholder feedback is considered.

Feedback and learning come from the Iteration Review, the daily Inspection activities, and the Team Learning objectives. In addition, the Enhancement Workshop is used to create measurable performance improvements. These inputs are taken into the Context Assessment at the beginning of each iteration. Finally, the team adjusts their approach by taking their improvement decisions into the Mobilisation and Planning activity for the next iteration. In this way (as shown in Figure 3), Agile Lineout is a pipeline of consecutive activities that produce incremental performance improvement and a high-value solution.

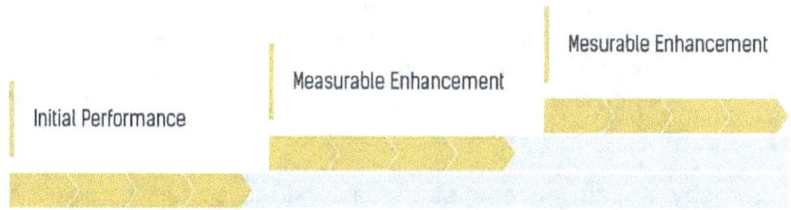

Figure 3 Incremental performance improvement

I am not suggesting that Agile Lineout will be a perfect fit for every situation; far from it, all models and frameworks have imperfections. However, because Agile Lineout is a "more complete framework" than Scrum, I have found that it:
- supports teams involved in strategic change.
- has benefits when coaching strategic change teams and new to agile software development teams.
- is scalable.
- can be used for programmes that combine strategic change and software activities.

Most Agile teams use a timeboxed approach to their activities. For example, a Scrum team has its Sprint cycle, and a Kanban team has a cadence representing the flow and rhythm of the Kanban meetings. The timebox is relatively short in software activities, two or three weeks. However, strategic change, transformational or strategic change activities require longer. I use a 30-day timebox and a rolling 30-60-90-day plan, as strategic change activities take longer to reach outcomes. Other business activities may require alternative timebox durations.

In Agile Lineout, these timeboxes are known as Iterations, and they should be as short as possible to allow for timely feedback and empirical control. In other words, the timeboxes should be a length that suits the activity outcome. However, all timeboxes should have activities short duration so the team can visualise its progress using a Kanban or a Scrum task board.

The Players

The delivery of strategic change is a team activity involving players with skill sets that combine to produce the desired solution outcome. This list of players in Agile Lineout is not intended to limit or restrict alterations teams may wish to make to improve performance. In certain situations, additional players may be needed and added to the team if agile principles are upheld. Such additional roles should not, for example, include team managers or others tasked to plan or control the self-managed, continuously improving team. These types of functions are often agile anti-patterns and should be avoided!

Agile Lineout focuses on defining the desired solution or outcome and how this is interpreted into action by the team. First, agile Lineout outlines the pivotal role of a Solution Owner. Then, it describes how Team members act to produce a strategic outcome or solution.

However, not all teams will have all of the skills needed to complete their goal, hence the need for extra-team support to provide additional expertise or part-time specialist knowledge. Agile Lineout describes the Supporting Cast role, which comprises specialists and supporters who are periodically engaged with the team. The role of the Team Coach is to facilitate and enable the Solution Owner, the Supporting Cast and the Team Members to deliver value.

The Solution Owner

Although I have described the role of the Solution Owner as if undertaken by a single person. Sometimes in strategic change activities, the solution or outcome definition is undertaken by a group of people known as a guiding coalition.

According to John Kotter, a guiding coalition is a group of individuals within an organisation who are the leaders of the change initiatives. They also have a social network which they can engage to facilitate the changes being made. See the Guiding Coalition section later in this chapter for further details.

Unlike software activities, strategic change initiatives may define the outcomes but lack detailed task definitions. Redefining the strategic goals defined as Outcomes and Key Results (OKRs) provides clarity of the intended outcome. During Mobilisation and Planning, the team takes these OKRs and, supported by the Solution Owner, creates the tasks they plan to undertake in the next Iteration. In this regard, the Solution Owner acts as a servant leader, focusing on the value of each Increment.

It is sometimes thought ideal for an agile team to access its customers directly. However, this is often not possible. Consequently, a Solution Owner often serves as the customer proxy for the team and works with the stakeholders, including other Solution Owners, to help define and prioritise the work in the backlog. From these descriptions, it is clear that the Solution Owner is the lynchpin of an agile team; it is the supercritical role.

Furthermore, as described by Etienne Wenger in his *boundary practice concept* (1998), the Solution Owner bridges the boundaries between the stakeholders in the broader organisation and the team members.

Solution Owners reflect the desires and expectations of stakeholders and convert these into outcomes and requirements. They act as a two-way conduit of information between stakeholders and the team.

Being a conduit has issues of trust. The development team may expect the protection of the Solution Owner. Yet, stakeholders may also see behaviour stopping them from getting what they need. These trust issues may be influenced by where the Solution Owner originated or resides in the organisation. This organisational heritage may also affect a Solution Owner's comfort in boundary spanning. According to Etienne Wenger, the Solution Owner uses boundary practice to maintain the connections between several organisational parts by *"addressing conflicts, reconciling perspectives, and finding resolutions."*

Because of these organisational tensions and politics involved in boundary spanning, I have suggested that a Guiding Coalition can function as a Solution Owner in strategic change activities.

The Agile Lineout system uses Big Room horizon-based planning to resolve conflicting goals or priorities and address some of the politics involved in strategic change. Big-room or quarterly planning is where stakeholders can be assembled to understand how the change initiative meets their needs or objectives. It is the dialogue between Stakeholders before and during big-room planning or when reviewing the outcome of an increment which tends to resolve the organisational tensions and politics.

To be effective in their conduit boundary-spanning role, Solution Owners need to use two languages simultaneously. For example, when working with Stakeholders, they hold conversations using business terminology and phrases. On the other hand, when conversing with the agile team, they often need to revert to technical or transformational language. Consequently, Solution Owners need to be skilled in building trust and creating purposeful relationships to facilitate their activities. Because of this trust-building and balancing role, a Solution Owner is sometimes required even when an initiative uses a Guiding Coalition.

The Solution Owner must therefore have a clear sense of the business benefits of the solution. In addition, they require knowledge and the ability to help the team create deliverables, activities, and tasks in the backlog, establishing a clear path between the work needed and the outcomes defined by the OKRs.

Working with team members, the Solution Owner facilitates the creation of the roadmap of incremental deliveries. A Solution Owner can enhance or reduce team effectiveness in many actions. For example, they need to define work items of an appropriate size so the team may consume them to manage the flow and speed of solution delivery. In strategic change activities, the Solution Owner must exercise diligence to ensure that all team activities help them progress towards their strategic goal. In addition, scheduling highly beneficial features early in the delivery timeline increases the rate of value creation.

A Solution Owner may not push the team to go faster except by creating smaller deliverables. The Team sets the delivery pace based on capacity. It is unacceptable to overload the team simply because the activity is behind. When an initiative takes longer than expected, a Solution Owner should address the scope using Kanban's capacity and flow data to determine the work that can be completed within the requested time.

Once the outcomes, increments, and deliverables are defined, the Solution Owner must be satisfied that the work items are completed as intended. This verification activity is critical to the speed of solution delivery and the quality of the outcome. Verification may, at its simplest, involve testing; however, more complex means of proof may be needed in cultural change activities.

A Solution Owner will be required at the programme level when big teams or programmes are involved. This overarching role works with the solution priorities to maintain the delivery and integration integrity of the features, components, or activities that each team is completing.

It is also the responsibility of the Solution Owner to provide progression data to the team. For example, suppose a strategic change activity was established to improve customer satisfaction. In that case, the Solution Owner could make the current measure of satisfaction levels, such as NPS, available to the team. Then, as the team delivers increments into the market, the impact of their efforts can be monitored, and they can pivot if needed. Similarly, progression metrics in a merger or acquisition could illustrate progress as functions are merged or processes and systems harmonised.

The Solution Owner's role is to maximise the value created by a team. However, they also have a role in analysing when the cost of delivering a feature exceeds the anticipated business value. In this regard, the Solution Owner can reduce waste by stopping activity based on investment economics.

Team member

The team members are collectively responsible for producing a high-quality solution or business outcome for the stakeholders, collaborating to achieve the Outcomes and Key Results. They also commit as a group when setting the Increment Goal in the Mobilisation and Planning activities. They use their collective abilities, energy, and intelligence to assess their context and achieve the desired outcomes.

In *Why drink is the secret to humanity's success*, Robin Dunbar suggested that individuals invest their time in creating relationships with up to fifteen people. Therefore, while an ideal Agile Lineout team may be constructed with up to fifteen players, this number may be extended when more people or specialist skills are needed. The key is to keep the number of people small enough that they can become an effective team. This limitation means that several teams may need to be created for significant activities requiring large numbers of people.

In an ideal world, the individuals in the core team will have the necessary core to create the strategic change or solution. However, that may not always be possible, necessitating requirements for the skills of supporting cast members, possibly as travellers or ambassadors.

Team coach

An essential principle of agile is that the team is self-managing and self-organising. Therefore, the Team Coach's role is to guide the team to be more effective, and it is primarily a role of enablement and facilitation. With Agile Lineout's servant-leading stance, the Team Coach role will continuously challenge the team to seek improvements.

The Team Coach focuses on delivering valuable work items by enabling the Solution Owner and facilitating the team Mobilisation and Planning activities.

Using the Inspection meetings, the Coach helps the team evaluate their progress and adjust their plans to achieve the Iteration Goal. They do this by analysing their performance and looking at WIP limits, cycle times, delays, and their conclusions about work to be completed.

Midway through the iteration, the Team Coach facilitates the Enhancement Workshop. This workshop is where the team evaluates their performance and metrics using tools such as lead time, cycle time and graphs such as burndown, velocity, and control charts. Next, the team considers their delivery value stream and seeks to identify improvement options. They then decide the changes they wish to make and how to alter their working methods. Finally, the Team Coach encourages the team to create and maintain their Improvement Backlog based on the outputs of the Enhancement Workshops.

Following the Enhancement Workshop, daily Inspection activity focuses on performance to completion. In addition, the team considers integrating components, expediting activities and completing work items necessary to achieve the Increment Goal.

When working with an Agile Programme or team of teams, several Team Coaches may be needed, and one of these will need to support the Agile Lineout at Scale activities.

Guiding Coalition

In strategic change activities, it is necessary to assemble a group of people with enough power to lead the change and encourage the organisation to work towards a common goal. Dr John Kotter calls this group of people a Guiding Coalition and defines them as individuals who are the social leaders of an organisation.

A crucial feature of a Guiding Coalition is its diversity. It should include individuals from across the organisation who can bring various perspectives, skills, experiences and a network of relationships. These networks can be called upon to enable and focus the strategic change activities.

The various backgrounds allow the team to see alternative solutions to challenges and encourage innovation. Depending on the nature of the strategic change, the Guiding Coalition may also require specific skills or backgrounds. Specialities often required include; change management, financial, legal, strategic planning, human resources, and technology.

Members of a Guiding Coalition do not all need to be senior people. An effective coalition has the right mix of individuals at different levels of the enterprise who display the following characteristics:

- Organisational knowledge,
- Specific expertise,
- Political credibility,
- Time for involvement,
- Functional leadership.

Recruiting people with these attributes can be challenging, as the individuals with these characteristics are often the ones the enterprise depends on most. Therefore, these guiding leaders must be empowered to make the radical long-term decisions a transformation demands. Their varied backgrounds, roles, and job titles, coupled with enthusiasm, will create momentum, resolving challenges to push the strategic change forward.

Supporting cast

The supporting cast comprises specialists who can be called upon to support the core team when needed. For example, such specialists could be lawyers, accountants, business architects, technical architects, or individuals with specialized domain knowledge such as regulation, robotics, or artificial intelligence.

When planning their work, the team will need to take a forward view of their need for supporting cast members so that they do not encounter delays or wasted activities. Work for the supporting cast should be included on the team task board to see progress, challenges, and dependencies.

Agile Programme Manager

When scaling Agile Lineout, an additional role is required to coordinate and synchronise deliveries from several teams. I have called this role the Agile Programme Manager. In SAFe, this role is known as a release train engineer.

The role and duties of the Agile Programme Manager include:
- **Managing and optimising** the delivery of solution value from multiple teams
- **Coordinating activities** to minimise delay and reduce waste.
- **Facilitating** the programme level Mobilisation and Planning activities
- **Supporting the programme level Solution Owners** as they decompose the solution and define work items for teams
- **Influencing** the programme-level quality plan, strategy, and integration activities
- **Acting as custodian** of the programme-level delivery and operational risk management activities and assessments
- **Providing an escalation path** for teams encountering impediments or constraints in their delivery activities.
- **Operating the programme** within the fiscal and delivery governance frameworks of the organisation.

Context Assessment

Assessing the context of the activity at the beginning of an Iteration is an explicit part of Mobilisation and Planning. Each activity element is evaluated, starting with information received during the Iteration Review, moving on to the delivery process, and considering adjustments to their team working.

In the Context Assessment, a team evaluates their current situation and adjusts their plans to reach the Outcomes and Key

Results or Solution Goal. The Context Assessment is linked to the Enhancement Workshop as a mechanism for continuous improvement. However, I do not see this as a second retrospection. Instead, I see it as a consolidated appraisal facilitated by the Team Coach. In this appraisal, the team identifies the changes needed for the next Iteration in readiness for planning.

Evaluating the solution context enables the team to decide if improvements or changes are required. For example, Stakeholder feedback during the Iteration Review may need the Solution Owner to alter the Increment Roadmap or revise the backlog. Alternatively, evaluating their progression metrics may cause the team to pivot or change their release plans.

A candid assessment of the delivery process may cause the team to alter their working methods. For example, measurable improvements suggested by the Enhancement Workshop must be considered for the next Iteration Plan. There may also be learning from the completion of the previous iteration, which means the delivery process needs to change. Finally, changes to the delivery process nearly always have a knock-on effect on estimates and team performance metrics, which need to be recognised in their plan.

The team working may also need to be adjusted because people have left or joined. They may require additional skills or changes in responsibilities to make them more effective. During the last Iteration or increment release, they may have found that their Working Agreement requires adjustment. In a programme, the organisational construct may require alteration to improve the big team's performance, to manage better dependencies or cross-team working. The Contextual Assessment of teamwork may alter the Working Agreement, put items on the Improvement Backlog, or place topics on the Team Coach's plan.

The Context Assessment is a quick stock take, not a lengthy separate review. It is an evaluation which serves as input to the mobilization and planning activity. The Context Assessment helps a team to adjust its plans for subsequent iterations. The Context

Assessment picks up threads from the Enhancement Workshop, the Learning Parking Lot and the Improvement Backlog.

The Solution Line

The Solution Line has four strategic steps repeated or refined for each iteration (see Figure 4). First, the Solution Goal is defined in terms of Outcomes and Key Results for the strategic change activity. Next, Benefits Mapping is used to break the solution goal into work items which are then sequenced to determine the Increment Roadmap. Finally, the Iteration Review provides the Guiding Coalition and the Team feedback.

These four strategies are built on the foundations of: solution quality, work item sizing, stakeholder feedback, and establishing the value of each increment.

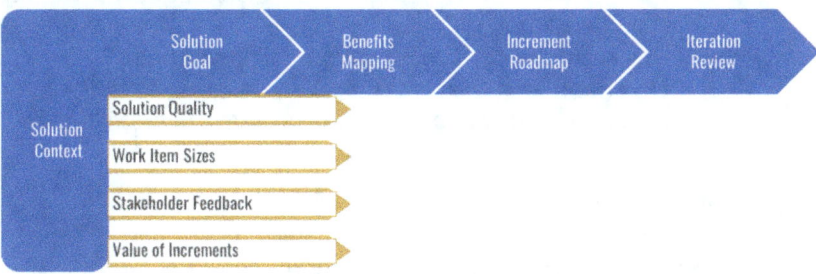

Figure 4 The Solution Line

The Solution Line is part of the Agile Lineout system of delivery. Therefore, while it is described here as a single element, the Solution Line is integrated and aligns with the Development and Team Behaviour Lines as a single delivery system.

Before beginning each increment, the strategic steps are preceded by an evaluation of the Solution Context.

Solution Context

Starting the Solution Line with an evaluation of the Solution Context enables a thorough preparation of the Backlog and gives the team clarity regarding the outcomes required.

In the case of a new activity, the evaluation of the Solution Context is used for initial sizing and scope definition of the outcome.

29

It may also be used to identify potential skills needed in the team. However, the Solution Context evaluation is used in an ongoing activity to understand stakeholder feedback or address quality issues discovered in the last Iteration.

The rigour of the Solution Context evaluation enables an assessment of the Increment Roadmap. It facilitates defining the next increment goal for the team. This contextual evaluation considers the following:

- Is the Solution Goal still relevant?
- Have any of the OKRs or Solution Goals been altered?
- Do the Progression Metrics or stakeholder feedback show clear progress towards the desired outcome or suggest a need to pivot?
- What was the feedback from the last Iteration Review?
- Has there been any regression of changes previously thought complete?
- Are corrective or consolidation actions needed?
- Do changes in the target market, competitive actions, or new product requirements necessitate the team to pivot?

Solution Goal

The Solution Goal describes the strategic goal or outcome needed from the team's activity. This Solution Goal may be defined as one or more Outcomes and Key Results (OKRs).

Vision statements are often collected from Stakeholders regarding the initiative's goal. Sometimes these are statements of aspiration intended to communicate a direction rather than a desired result.

Vision statements may be poorly constructed, contain assumptions or rely upon the interpretation of others to be complete. Such statements show how the team's activities align with the organisation's strategic mission or tactical needs. However, vision statements must be fully understood for progress measurement and delivery purposes.

OKRs provide a practical way to define the strategic or business goal accurately. The key results can be used to illustrate progress and the results delivered.

Based on stakeholder discussions, the Solution Owner sets the initiative's direction. At the outset, they identify the aspirations and benefits of the solution. Usually, in strategic change activities, this definition is in the form of Outcomes and Key Results (OKRs).

Programme Name: Impetus		
Programme Level Outcome	Next Quarter Outcome	Key Result
Place in Top Five in sector by market share	Improved customer satisfaction	• As measured by • Customer NPS:
Maintaining if not improving current profit levels	Cost saving from merged organisation	• As measured by • Reduced costs by:
Market leading customer care offering	Increased sale due to customer care programme	• As measured by • Increased revenue amount:
		• As measured by

Figure 5 Aligning OKRs

Figure 5 Aligning OKRs, illustrates how overall outcomes for a programme or a year are decomposed into the next feasible valuable delivery, with the key results as measures identified for the next quarter. The final column would have numbers, so, for example, improve NPS by 10 points.

OKRs describe the overarching goal of the strategic change initiative. However, they may also be used to explain the intermediate outcomes during incremental delivery. In this context, the Solution Goal is refined and broken down into actionable work items using Benefits Mapping. This mapping is the first step in Mobilisation and Planning. Then intermediate OKRs are used to define the Increment Roadmap. (See Figure 6).

This clear statement of the Solution Goal in OKRS serves as a call to action. It justifies each team member's investment in achieving the outcome. It also provides an initial definition of the scope of the activity.

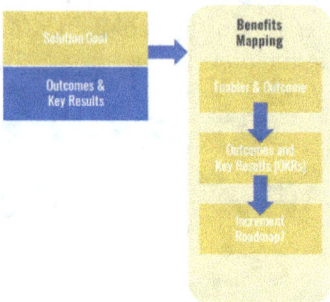

A clearly defined solution goal constantly reminds a team of the desired outcomes. Consequently, it should be prominently displayed on a big visual chart or agile radiator.

For completeness, the solution goal should eventually comprise inclusive and exclusive phrases to set

Figure 6 Linking OKRs to Benefits Mapping

boundaries for the initiative. However, the team may add or refine these as they progressively explore the scope of the strategic change outcomes.

The key results or Progression Metrics provide feedback enabling the evaluation of progress towards the desired outcome. In some instances, it may be necessary for a team to create the management information they need as part of the scope of their activity.

Where teams participate in an exploratory activity, the vision statement may describe a direction of travel rather than a specific outcome. In these circumstances, the team may use techniques such as design thinking to define their objective and use the Iteration Review for stakeholders to confirm their incremental definition of their objectives.

Benefits Mapping

Strategic change activities prepare a backlog by decomposing the Solution Goal using Benefits Mapping. Typically, backlog preparation is a team activity led by the Solution Owner and may involve stakeholders. Its purpose is to decompose the Solution Goal into constituent parts. Benefits Mapping visualises this construct and enables valuable increments to be defined.

By continually mapping the benefits of an initiative, a team breaks down the OKRs or Solution Goals into a list of things the project will either make or the outcomes it will deliver. This activity may use mapping techniques such as Benefits Mapping or agile modelling. These techniques aid the creation and ongoing refinement of the Backlog.

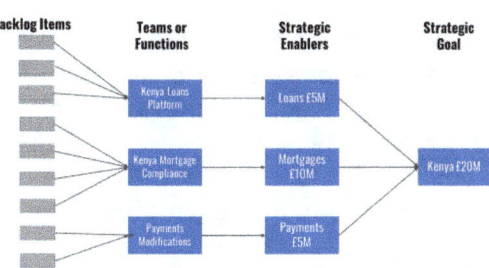

Figure 7 An example Benefits Map

As Key Results are accomplished, the Benefits Map is updated so that stakeholders and team members have a clear picture of the results achieved to date and those yet to be realised.

In the example shown in Figure 7, the strategic goal is to grow the Kenya business by twenty million pounds. The outcome is business growth, and the key result is by the measure of twenty million pounds. In addition, targets by product area have further refined the growth target; Loans by five million pounds, mortgages by ten million and payments by five million.

A collaborative benefits mapping exercise involving key stakeholders has shown that to increase the revenue from Loans by five million, the use of the platform needs to be expanded. This work could be a mixture of marketing and software enablement, but the relevant Backlog Items will be tasks on the left of the map. Similar analysis and expansion activity has resulted in the need to improve mortgage compliance and modify the payments offering. Once these modifications are identified, they populate the backlog items to the left of the outcomes and measures. As the OKRs are broken down into work items, the key results are also decomposed so that each activity placed on the backlog has a measurable outcome or completion criteria.

When creating the backlog, the Solution Owner and the Team also consider risk. In strategic change activities, two types of risk are typically managed: namely, development risks and operational risks.

Development risks anticipate how the delivery activity may be potentially interrupted or challenged. Therefore, development risk management seeks to create plans which nullify these potential hazards and increase the probability of success.

In Agile Lineout, development risk management follows the traditional project management approach of identification, quantification, resolution and monitoring. Risk mitigation activities are placed in the Solution Backlog alongside the business requirements. Teams monitor development risk levels during Mobilisation and Planning, Inspection activities and Iteration Review.

On the other hand, operational risk management seeks to manage adverse impacts from inadequate or failed internal processes, people and systems, or potentially from external events such as fire or flood. Operational risks may affect client satisfaction, an organisation's reputation, its relationship with its stakeholders, or shareholder value.

Operational risk levels are often altered when undertaking significant organisational or strategic changes. In operational risk management, risks are identified, risk owners are established, risk levels assessed, risk indicator levels are set, and controls are implemented.

The outcome of operational risk management is that the risk level associated with operating the business falls within the risk appetite of its management and stakeholders. Therefore, Solution Owners must consider operational risk assessments as an integral part of the solution delivery process. This level of rigour is critical where incremental or partial deliveries may expose temporary operational risks which do not appear in the final solution. If temporary operational risks are exposed, the team must implement temporary controls and key risk indicators in their development activities.

For compliance reasons, Team Coaches may need to ensure that any individual's responsibility for risk management is made clear in the Working Agreement.

Preparing the Solution Backlog as a group activity using mapping and risk management techniques ensures that all concerned agree on the components needed to reach the Solution Goal. In addition, during this activity, assumptions are exposed. The Team may explore these assumptions and challenge these if necessary. However, the outcome is that understanding increases, expectations are aligned, and interim goals are identified.

Once the initiative is underway, backlog refinement and mapping benefits become an almost continual activity where the Solution Owner and team members review and prioritise the work yet to be completed.

The team may remove, rank, or prioritise the work items in this activity. The priority determination should be based on data potentially from the Progression Metrics or stakeholder feedback. Alternatively, the priority could be set by input in the latest Iteration Review. Or could be resource-focused based on the remaining budget or time. Placing the must-do items at the top of the backlog will ensure they are completed. Finally, the Solution Owner and team ensure that work items at the top of the backlog are ready to be worked upon.

Backlog refinement occurs regularly in each iteration and may be a scheduled event or an ongoing activity. Some of the activities that arise during this refinement of the Solution Backlog include:

- **Removing work items** that no longer appear relevant.
- **Creating new work items** in response to newly discovered needs.
- **Adding work items** that mitigate or reduce risk levels and introduce risk controls or key risk indicators.
- **Re-assessing the relative** priority of Increments and work items.
- **Estimating work items** or evaluating the estimates of items previously assessed.
- **Splitting work items** that are too large to be completed in an upcoming iteration.

The appropriate sizing of work items is a critical element of establishing an efficient team and monitoring progress and is a foundational focus in Agile Lineout.

Increment roadmap

Agile activities using an incremental delivery process. An incremental approach is one by which an outcome is built and delivered in pieces representing a subset of the Solution Goal. The sequence of deliveries is known as an Increment Roadmap.

The Increment may be either small or large, depending upon the context of the activity. The critical factor is that each Increment must be valuable and viable, which means that it can be used and not expose the organisation to any potentially unacceptable risks. Sometimes it is necessary to visualise or estimate the anticipated impact of actions to check the viability of an increment.

The Solution Owner, with team members, is responsible for breaking the solution into viable increments. The sequence of increments is the road map. Each increment has an Increment Goal, a statement of the intended value created. The Increment Goal may be in the format of an OKR or a change in business metrics. The Increment Roadmap becomes a vehicle for communication with stakeholders and reminds the team of the path to their outcome.

In assessing viability, the Solution Owner must ensure that each implementation of an increment does not significantly alter the organisational risk profile. The relevant controls are introduced to mitigate the identified risk if it does. In regulated environments, viability is a necessity. A coach may need to remind the Solution Owner of an increment's compliance or operational risk implications.

The increments are iterative. The team plan for the outcome from one increment will be improved upon in subsequent increments until the solution goal is realized.

Iteration review

A review with stakeholders can happen at any time but, as a minimum, should be after each iteration's completion. This appraisal is especially important for strategic change activities. Because change, particularly cultural change activities, often progress as a "drunken walk", moving from activity to activity and staggering a little, returning to refix something already thought fixed. Therefore, the need for course correction or adjustment with cultural change activities is a frequent occurrence.

In the Iteration Review, the Solution Owner orchestrates an event to evaluate the status and illustrate progress towards the Increment Goal. Some Iterations will complete an Increment, in which case a review of the entire increment replaces the Iteration Review. The Iteration Review is essential as feedback provides the basis for empirical control.

Often with cultural change activities testing or checking if the actions have had the desired result is a complex but necessary step. Sometimes it is required to use a World-café approach. The World-café asks powerful questions to solicit feedback on situations or improvements. Dialogic Organisational Development suggests that dialogues manifest the culture of a firm. Therefore, it is by using dialogue that culture can be changed. World-café interventions can indicate the next feasible right step and test the impact of previously taken actions.

The Iteration Review allows single-loop learning, meaning the team understands feedback on the solution produced to date. However, Agile Lineout combines the Iteration Review with evaluating the Solution Context. This combination allows double-loop learning. Such evaluation may cause the alteration of solution scope, change of development processes, and reassessment of assumptions. During the Iteration Review, a team may also learn that it needs to pivot or cancel further development.

The Iteration Review is a meeting orchestrated by the Solution Owner in which the team and relevant stakeholders appraise the OKRs. There will be a demonstration of achievements, the changes to the business or the presentation of metrics illustrating the improvements made. Or, in the case of any software development, a demonstration of a working tested product.

However, the Iteration Review is more than simply a demonstration; it is a verbal equivalent of a written status report. An event with stakeholders to consider the outputs of the iteration A typical format of an Iteration Review is as follows:
- A comparison between the planned and actual Outcomes for the Iteration.
- An evaluation of Key Results or Progression Metrics to illustrate the impact of the strategic change activities.
- A summary of current delivery metrics, which may or may not include financial data.
- A review of the current delivery and operational risk profiles.
- The communications plan and updated stakeholder map should be discussed in strategic change activities.

A detailed review of Progression metrics will illustrate how the activity is advancing towards the overall OKRs and the Solution Goal. Adverse progression metrics may cause the team and Solution Owner to consider a pivot or a change in scope or plan. Or they may indicate that the activity should be halted.

Foundational items

In addition to the above strategies, four foundational items are of continual focus for team members in the Solution Line. These are:
- The value of increments.
- Work item sizing,
- Solution quality, and
- Stakeholder feedback.

Each of these items is described in detail below.

Value of increments

Establishing the value of each increment is essential as the Solution Owner and Team prioritise work based on the anticipated value being created. Creating value and avoiding non-value-adding items is the essence of agile, especially in a strategic change where the linkage between activity and outcome may be tenuous.

When coaching, I find conversations with Solution Owners often turn to business value and viability topics. In particular, I am frequently asked how to prioritise one feature or story over another. The Benefits Mapping or Agile Modelling techniques described earlier provides increased clarity over an increment's content.

The apportionment of value doesn't have to be accurate but near enough to guide prioritisation. For initiatives with no monetary value, a percentage can be used by setting the amount at a fictitious one million or a number valid in the scale of the organisational portfolio.

Sometimes the value may be obscured. For example, a compliance activity may mean the organisation can stay in business, but calculating this value in monetary terms is pointless. In these obscure instances, it may be necessary to introduce synthetic data, such as value points, to help a team prioritize work.

Work item sizes

To create solutions quickly, we need work items to flow through the delivery pipeline or system without being blocked or slowed down. Breaking solutions into smaller work items enables their rapid completion, increasing the opportunity for feedback and facilitating an early creation of value. An agile team constantly asks if we can break this work item down and deliver it faster. Flow metrics can be used to guide a team on whether they are delivering value quicker or not.

Some Solution Owners have difficulty breaking features down, vertical slicing and defining the outcome as a series of increments. Yet, these actions directly impact how effectively the team can deliver and are, therefore, essential concepts to master,

In addition, some novice Solution Owners find the concepts of a minimum valuable increment or minimum viable product alien since they previously, using traditional techniques, expected the whole list of requirements to be delivered simultaneously at the end of the project.

The appropriate size of work items will depend upon the nature of the activity. For example, in software projects, work items may be optimally sized around a day's work, whereas an organisational change activity may require work items of a longer duration. In addition, some items, such as those externally procured or deliverables with a long lead time, may have durations outside normal tolerances. However, creating smaller-sized work items benefits the development process by significantly increasing the speed of delivery.

Work items should be sized so they can be produced and evaluated as being complete in a single iteration.

Solution quality

There is no point in an agile team quickly producing items unsuitable for purpose! Consequently, the ethos used in agile is known as shift-left, which engenders a right-first-time culture. By shift-left, I mean in strategic change activities using OKRs to define the expected quality of work items as early as possible in the delivery process or using agile techniques such as test or behaviour-driven requirements.

The Solution Owner plays an often-forgotten role in the quality of the outcome. For example, in Scrum, the Solution Owner is the only team member empowered to accept work as completed. I have not been this prescriptive in Agile Lineout, but I would expect a Solution Owner to have a significant say in deciding if a work item is complete.

As the Iteration Goal is often stated in terms of a working tested solution, this role requires the Solution Owner to be satisfied that the team's quality strategy is appropriate for the solution being developed. Additionally, the Backlog must include any continuous integration elements and end-to-end quality assurance activities, items that Solution Owners sometimes forget.

I have often seen agile teams struggle to understand if they have produced the right thing. Novice Solution Owners often make vague requirements and expect the team to understand the statements' nuances, assumptions, and implications. Defining the key results as acceptance criteria alongside each objective or work item dramatically reduces confusion. It tells the team you will have got there when XYZ is true. If a Solution Owner cannot define the key result or XYZ test, they probably will admit that they don't know enough about the work item!

So how does a team ensure the quality of the solution they are building? In strategic change activities, OKRs provide the as-measured-by statements that clarify the outcomes and offer targets that can be tested. The testing or quality assurance aspect comes from collecting the KR data.

This test-driven requirement approach (defining the OKRs for strategic change upfront) clarifies the goals for change and software activities without resorting to detailing the requirements at the beginning of the initiative. Creating test-driven requirements is as simple as defining how the team will know they have produced the desired result. Test-driven approaches simplify or clarify outcomes and shorten the time needed for requirements definition.

Stakeholder feedback

Short feedback loops in agile are essential to efficient delivery as they ensure the team remains on the path towards their solution goal.

Obtaining good Stakeholder feedback in an organisation newly adopting agile is frequently challenging. This situation requires Team Behaviour that encourages participation and often involves explaining why regular and early feedback is essential to the agile delivery system.

In some cases, the actual stakeholders of a team are external customers. Therefore, it is difficult for a team to get customer feedback on each iteration. Feedback in such situations often comes from stakeholder proxies such as product management or the marketing team.

For internal initiatives, the complexity may be generated by the number of stakeholders and their differing expectations from the team.

Stakeholder feedback ensures that the team has access to meaningful, timely, and appropriate information and that the solution developed meets or exceeds the business expectations. The stakeholder feedback is collected by the Solution Owner, who uses the data to refine the Solution Backlog, prioritise work or pivot the delivery system.

Artefacts in the Solution Line

The Solution Line focuses on the Solution Owner's actions and the stakeholders external to the team. It emphasises building the right solution in increments that make business sense and quickly create value.

The following table (Figure 7) shows how Agile Lineout artefacts support the Solution Line. The table is intended to illustrate primary usage, but a team could choose any artefact to be used more extensively than I have suggested.

The Solution Line uses several artefacts to illustrate what is being developed. However, some artefacts exist because the team needs to understand and then decompose the Solution Goal into work items and deliver them appropriately.

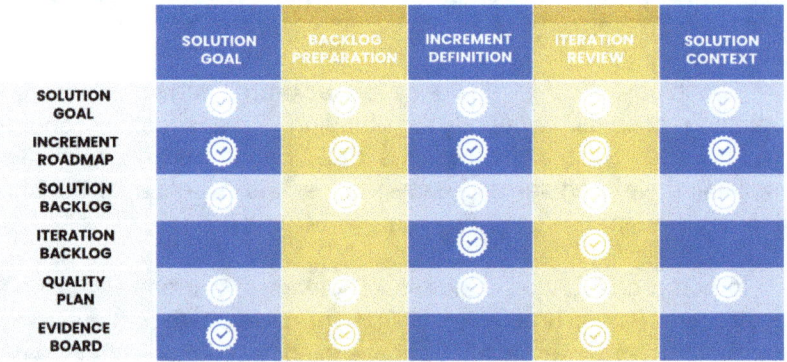

Figure 8 Artefact support for the Solution Line

Detailed descriptions of each Agile Lineout Artefact appear later in this book. The emboldened ticks show the strategy which creates or maintains the artefact. The shadowed ticks show artefacts that are considered or serve as inputs to the Solution Line strategic step.

The Development Line

The Development Line is the second element in Agile Lineout. It is concerned with the process of turning the Solution Goal into reality. In addition, the Development Line helps the team establish an efficient means of production or value stream.

As shown in Figure 9, the Development Line has four strategic steps repeated or refined in each iteration. These steps are; Mobilisation and Planning, Inspecting Performance, the Enhancement Workshop and Inspecting Completion. In addition, an evaluation of the Delivery Context as part of the Context Assessment precedes these steps.

These four strategies are built on focus foundations of establishing Ways-of-working, Efficiency, Quality Process, and Performance Improvement.

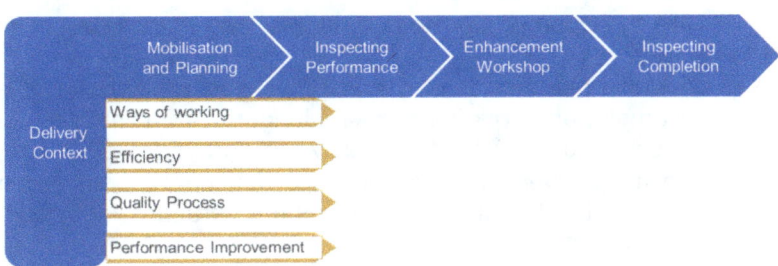

Figure 9 The Development Line

As previously described, the Development Line is part of the complete delivery system. Therefore, while it is explained as a single element, the Development Line is integrated, aligns, and is interdependent with the Solution and Team Behaviour Lines as a solution delivery system.

The Delivery Context

When starting a new activity, the Delivery Context evaluation will be a first stab at answering the question, "How will we develop this solution," e.g. What processes, tools, and techniques will we need?

When starting an enhancement activity to an existing Solution, product or service, evaluating the Delivery Context explores the previous development process. It enables a team to identify the development value stream used in the past and decide what they may need to change or adapt to reach their new Solution Goal.

With an existing activity having just completed an Iteration, the team decides how their production process needs to change based on what they have learned. For example, the Iteration Review may have identified:

- A high level of escaped defects.
- Expected outcomes being missed or key results different to those targeted.
- They may have found a need for more requirements modelling.
- Maybe a new tool or an additional process is required.

Therefore, the evaluation of the Delivery Context considers the development process, including:

- Are the appropriate methods, tools, and techniques being used?
- Are the plans and estimates proving to be realistic? For example, are work items taking longer or remaining on the Backlog due to the unavailability of time or people?
- Do the delivery metrics show the continual advancement towards the Solution Goal? Does the Iteration plan require adjustment?
- Are the experiments or improvements suggested in the Enhancement Workshop creating measurable performance improvements?
- Are improvements suggested in the Enhancement Workshop not yet implemented by the team?
- When Inspecting, does the team ask itself whether it will reach the Increment or Iteration Goal?
- Is the team adhering to its quality plan? Are the testing activities revealing poor quality, or does quality need to improve? If so, what changes do they need to make?

The Delivery Context evaluation may also indicate a need for a change if the team membership has altered in a significant way.

Mobilisation and planning

Mobilisation and Planning is the team's development line strategy to start the next iteration. Some people have asked why I placed the word mobilisation before planning. The answer is that a team needs to explore the scope of the activity before any planning can commence. This statement illustrates how the Solution Line interacts with the development line as part of the delivery system and explains why I described the Solution Line first in this booklet.

Many strategic change activities take time to realise their intended results. Consequently, the overall goal is broken down into sub-goals defined by OKRs which can be achieved during the next quarter. Once agreed, these OKRs can be decomposed into actionable work items either progressively or as a single backlog refinement activity. However, by clearly setting the direction in big-room or quarterly planning, the Solution Owner or Guiding Coalition can orientate the team's attention, engage their talents, and encourage action.

In big-room planning, team members collaborate with their stakeholders to create the desired outcomes and key results for the next quarter. In doing this, the emotional and intellectual energy in the team is fashioned by emphasising the strategic imperative of the strategic changes they are to produce.

Initially, the big-room activity creates a roadmap of the increments for the complete initiative based on a series of horizon statements defined by OKRs. In the big-room planning event, a team will first outline its current status, review the progression metrics, and evaluate risks, before agreeing on the OKRs for the next horizon. When defining the outcomes, the Benefits Map will be used as a reference and to guide the planned next steps.

As shown in Figure 10, Mobilisation and Planning is based on the Benefits Map. The team takes work from the Solution Backlog to create the Increment Goal. It then breaks down the work required into items sized so that each may be completed in a single iteration, thus making the Iteration Backlog.

When starting a new activity, a process tailoring workshop is needed. This workshop, facilitated by the Team Coach, enables the

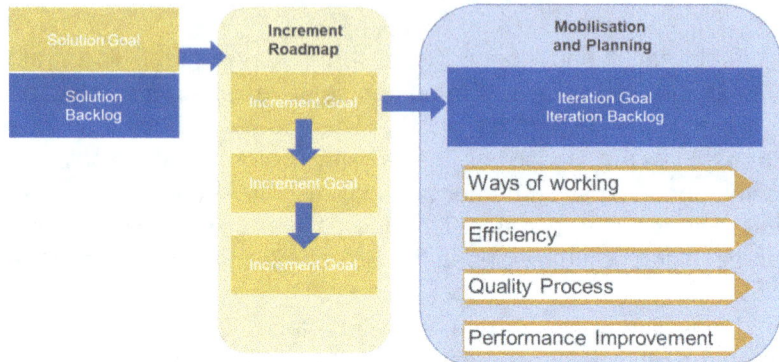

Figure 10 Mobilisation and Planning

team to decide on the tools and techniques required to create the Solution. These ways of working are typically documented in the Working Agreement. They are heavily influenced by the activity ecosystem provided by the organisation. However, sometimes a solution may require a team to request additional or new tools and techniques.

Teams sometimes ask if they should use Kanban with Agile Lineout. My answer is yes. Kanban aims to give team members enough work to operate at capacity consistently. Teams that practice Kanban benefit from flexible planning, clearer focus, and total transparency because whatever's on the board is the top priority and the items on which the team is working. Kanban is also great for operational teams focused on continuous delivery or teams with frequently changing priorities.

Most agile teams use Jira, a tool from Atlassian or Azure Dev Ops from Microsoft, and both tools are designed to help teams manage work. For example, both tools support the development of solutions in iterations. In addition, both have visual task boards and collect metrics to help a team control its work and productivity.

For teams who wish to use Kanban, Jira has a function called Kanplan. Kanplan facilitates backlog refinement in the way advocated by Agile Lineout.

Software Kanban boards traditionally don't have any backlog functionality. Instead, Solution Owners and Team Members use work items in the first column to plan. The Kanplan feature in Jira introduces a backlog view to enable teams to refine, prioritise, and then move work through the workflow. For a more substantial alignment with Agile Lineout, the story mapping feature may be combined with the Benefits Map to give a complete picture. Strategic change teams may therefore find the Kanplan function in Jira helpful.

Kanban allows planning without estimating each work item. Planning without estimates may seem an anathema (something people hate), but in strategic change activities, it may be necessary.

First, however, we must ask ourselves why we are estimating. I expect the answer is to be predictive or tell stakeholders the dates for deliverables. However, with many strategic change activities, the outcome is often more important than the ability to predict a date.

Consider a software project that is overrunning. In this case, the team has two options: change the estimates or squeeze the work into the existing timeline. In both cases, the software team is on a downward spiral to finish work and hit the deadline. However, with organisational or cultural change activities, working faster may not be possible and increasing speed may result in irreparable damage to the intended outcome. So, the essential element for strategic change is sometimes the definition and realisation of the outcome, not necessarily completion within the timeline estimate.

With cultural change activities, traditional estimation approaches may be challenging. Some strategic change activities may be new to an organisation. Work items may also be encountered in which the team members have no experience. So how does a team estimate?

Kanban allows teams to collect average-time-to-completion data so that teams may anticipate when work may be done. This is where dividing work into small chunks and adjusting each iteration pays dividends. This is because a team discovers how long things really take compared to what they anticipate.

Based on recent delivery performance, Kanban averages the lead time for similar types of work as it passes through the development workflow. It is, therefore, essential that the workflow should cover all aspects of the development from start to finish. The lead time can then be used to plan, assess progress, and provide stakeholders with the anticipated completion time.

As previously stated, the duration of each iteration should be as short as possible to allow for timely feedback and empirical control. In other words, the timeboxes should be as short as possible but long enough to allow meaningful progress towards the Solution Goal. Agile Lineout can flex the length of an Iteration based on the viability and the work content needed to complete the Increment.

The team take the work items needed to create the Increment Goal from the Solution Backlog. However, the work required to reach the next Increment Goal may take more than one iteration. Therefore the team create an Iteration Goal as an interim objective on the way to the eventual Increment outcome.

In non-software activities, particularly those involving organisational or cultural change, a team should create a plan based on the concept of basecamps and expeditions.

The basecamp model uses the achievement of the Increment Goal as a point to reflect and take stock before pushing forward. On reaching their interim goal, a team may need to consolidate their achievement or stabilise the situation before pushing on to the next Increment. Frequently in strategic change activities, consolidating or stabilising actions cannot be planned in detail in advance as they are often reactive to the situations encountered. However, allowances should be made for these activities before commencing the expedition towards the next Incremental Goal or basecamp.

Inspecting performance

Once the team has completed their Mobilisation and Planning activities, they start the work. This is when the inspecting elements of Agile Lineout come into play.

Agile teams traditionally use a daily short time-boxed meeting to synchronise their work, inspect progress and adapt their plans. I have retained the regular short inspection meetings in Agile Lineout but made two alterations. In the first half of the Iteration, the inspection meeting emphasises performance. Whereas in the second half of the Interation, the focus switches to completing the Iteration Goal. These alterations are designed to encourage a team to look beyond an individual's work to the overall team situation in achieving the Increment Goal.

In switching the focus of the inspection meeting, I hope to make fulfilling their goals a regular occurrence. In addition, I have advocated using metrics such as the number of items in progress and cycle times to assist teams with their performance diagnosis and improvement plans.

Depending on the speed with which the team completes work items, they may decide that a daily Inspection activity is unnecessary. However, the more frequently they Inspect, the more predictable their delivery system will be.

Using a Kanban board creates the visual transparency needed when Inspecting Performance. Visualising work enables a team to identify problem areas and suggest solutions quickly. In addition, the use of work-in-progress limits focuses the team on completing activities rather than having many activities in-flight at any one time.

In this Inspecting Performance activity, the team may consider work items regarding lead time, estimated value, quality of deliverables, impediments, and the development processes used. The team also identifies any work items they believe need attention. Then, outside the Inspection meeting, it organises pairs, huddles, or swarming to create an action plan to resolve their challenges.

During Inspecting Performance, the Team Coach may need to draw attention to the metrics and trends identified in the data and facilitate the conversations during and outside the meeting to resolve issues. Failing to challenge a team in this way will reduce the intended rigour of Agile Lineout.

During the Inspection activities, the team must candidly evaluate how many items they have simultaneously in progress. In the

Working Agreement, they may have documented their attitude towards work-in-process (WIP) limits. Initially, a team may not know their process or which process step is the slowest. In this case, the team must do some work and observe. When they have identified their slowest step, they can set WIP limits by multiplying the number of items by the number of team members doing that step plus a buffer of fifty per cent. The buffer ensures that people doing the slowest step never run out of work to do. However, WIP limits are never set in stone and should be regularly reviewed and adjusted by inspection.

Enhancement workshop

A core theme in Agile is continuous improvement. Therefore, teams should seek better working methods in Agile Lineout to optimise their delivery system. Accordingly, an essential strategy in the Development line is the Enhancement Workshop.

The Enhancement Workshop is a facilitated opportunity for a team to reflect on all aspects of their delivery system and seek measurable improvements. In Agile Lineout, the Team Coach organises and facilitates the Enhancement Workshop. It should happen as close to the middle of the Iteration as possible.

An Enhancement Workshop in the middle of the iteration allows course correction enabling a team to become more predictable. With a workshop midpoint of the iteration, the outcome is alterable rather than a fact at the end. Instead of the team reflecting on the question, "How did the Iteration go?" the team now asks itself, "How is it going?" This question encourages performance enhancement as they correct their path towards the Iteration Goal. I have also proposed two further questions for the Team Coach: "How can we increase quality?" And "How can we be more effective?" These questions directly align with the purpose of the retrospective in Scrum and are action orientated.

The Enhancement Workshop should use metrics such as a Control Chart to show the average lead time for the team's work. If a team tries to "game" this metric, it will encourage them to identify smaller work items that can deliver precisely the desired behaviour faster.

The essential outcome of the Enhancement Workshop is a series of measurable improvements. These improvements should use objectives and key results (OKR) approach. The goal of the Enhancement Workshop is for the team to identify bottlenecks and improve their overall effectiveness creating the ability to build solutions faster with more predictability.

Inspecting completion

Following the Enhancement Workshop, the daily inspection activity takes an end-game flavour.

Inspecting Completion is where the value defined by the OKRs comes into play. In my experience, many teams try to do more work than they can. In strategic change initiatives, this situation is exacerbated by encountering unforeseen work. For example, a change the team thought would go smoothly provokes a political or emotional reaction. Additional communications, workshops or activities are therefore needed to quench the fire. The extra activity may mean that work the team anticipated could be completed in this Iteration cannot be done, and reprioritisation is necessary. During the Inspecting Completion activity based on the OKRs, a team can decide to pursue the completion of one work item while delaying another.

In Inspecting Completion, the team, in addition to their Kanban, may use a cumulative flow diagram or burndown chart to determine the current status and plan corrective actions. In this way, the team positively adjusts its work plan to achieve the Iteration Goal.

Foundational items

While the above four sections describe the strategies for the development element of Agile Lineout, the following four foundational items should be a continual focus of the Team Coach and Team. These are:
- Establishing the ways of working.
- Efficiency.
- The quality process, and

- Improving their performance.

Ways of working

Each team must decide how it will produce the Solution, how value will be created, and what tools and techniques will be used. In Agile Lineout, these are known as the Ways of working.

A Process Tailoring Workshop defines Ways of working at the beginning of an activity. This workshop is part of Mobilisation and Planning, and the output is documented in the Team's Working Agreement.

As the Process Tailoring Workshop may be one of the first times a new team meets and works together, there is also an opportunity to use this activity for the first steps in Team Building.

However, expecting a new team to decide all the processes they require in one session is unwise. Instead, it is better to let the ways of working evolve over the first two or three iterations. Furthermore, a Process Tailoring Workshop lasting more than ninety minutes should also be avoided. Instead, the Team Coach should apply the just enough principle, deciding only on immediate-use process elements.

Depending on the Solution Context, the processes that most teams need are the following:
- **The exploring scope process** defines how a team identifies and refines its Solution Goal and defines the OKRs. This process sets the parameters of the endeavour.
- **The Increment planning process** defines how a team breaks down the solution into viable increments creating the Solution Roadmap. This process also determines how interim OKRs define each increment and how the value of each increment is assessed
- **The development process** defines how the solution will be created. It also outlines the necessary tools

and techniques and where artefacts will be stored and shared

- Addressing changing Stakeholder needs defines how the **change control process** operates, the approvals needed and who is involved in agreeing on directional alterations
- **The quality assurance process** defines the strategies and collects the data that establish how the team know the appropriate outcomes are being created.
- **Controlling and coordinating activities** defines how the team will work together, work with other teams and apply controls to their actions. Coordination is of particular importance for programmes or teams of teams.
- **Addressing risk** outlines how operational and delivery risks will be identified, quantified, mitigated, and monitored.
- **Collecting Progression Metrics** or Key Results is a process that defines how the team will know if their activities result in the appropriate business outcome.
- The **Governance and Reporting processes** detail how data regarding the activity will be provided to the broader organisation.
- The **Deployment process** concerns how Increments will reach the intended customer.
- The **Change Readiness Assessment** outlines how ready the customer is to accept and use the change.
- The **Communications and Stakeholder Management process** provides change enablement information to stakeholders meeting their communication needs.
- The **Continuous Improvement process** concerns how the team plan to improve its performance.

Some of these processes may be defined as part of the agile delivery toolkit at the organisational level. However, alterations may be needed to the standards based on the context of a specific initiative. For traceability reasons, any modifications to standards and peculiar team processes should be documented in the Team Working Agreement and subject to continuous improvement from the Enhancement Workshop, Inspection and Team Learning strategies.

Therefore, the team's work will depend upon combining organisational, product and delivery capabilities and those specific to a specific initiative. The team's organisation may also provide tools, processes, and governance to support the team. However, these may not be enough. For example:

- The initiative may require a higher-than-typical quality solution.
- The team may be developing a prototype where a market reaction is sought.
- The market conditions may significantly reduce the average time to market.
- The Solution Goal may be an organisational change or capability development that has not been undertaken before.
- The product may be technically complex, have components sourced from suppliers, or need certain security elements.

A Process Tailoring Workshop will evaluate what is provided in the light of their Solution Goal and decide if additions or changes are required.

Before establishing a new team, a Team Coach should explore these elements and decide on the toolkit, techniques and processes the Team will likely need and the organisation provides. In addition, the Team Coach should thoroughly understand the organisational governance mechanisms and how these should be applied to their activity.

A crucial coaching principle is that there are no best practices but poor and better choices based on the team context. However, not everything should be developed uniquely and anything which is developed needs to provide the data or controls needed at the enterprise level.

Efficiency

Measures of efficiency are the metrics that illustrate the team's ability to deliver. However, they may also show how enjoyable it is to work as a team and how predictable they are as a delivery system.

The focus of a Team Coach is to improve the team's efficiency, which often means evolving their ways of working and identifying potential improvements. Yet, it isn't enough for the team to identify possible improvements. They must implement them!

Simply letting people manage themselves may not improve delivery performance; instead, we must provide the data which enables them to control their workflow or value stream. Every team will have a unique value stream. However, the team may also be part of a more comprehensive development value stream comprising several teams. In monitoring, managing and optimizing these value streams, efficiency is created. Using lean principles, we learn to avoid handoffs and delays. A Kanban board will quickly identify where any constraints or bottlenecks lie, presenting opportunities for efficiency gain.

When evaluating options for improvement, a Team Coach should first encourage the Team to establish how they will measure a successful outcome. For example, if the Team thought an improvement would reduce delivery time for an activity, establishing and monitoring a cycle time in that area could be appropriate.

The Enhancement Workshop should consider the improvement results, but the team may also wish to assess the effects during their daily Inspection activities and in the Enhancement Workshop.

Quality process

Quality is more than just testing. Quality assurance is how the team establishes that the right solution is being created. In strategic change activities monitoring quality may be as simple as collecting an analysis of the OKR key result information.

Agile quality assurance happens throughout the activity instead of being the subject of a phase, as with waterfall projects. The definition of the Solution Goal in terms of OKRs and interim OKRs is the start of quality assurance by clarifying precisely the outcome required. In agile initiatives, the whole team is responsible for delivering a quality solution. Having stakeholders define the outcome and allowing the team to determine how those outcomes will be measured (the KRs) ensures clarity and collaboration.

Furthermore, providing outcomes in iterations and incrementally releasing them to customers requires careful planning and quality control if a team's or organisation's reputation is not going to be damaged. Agile Quality assurance uses a risk-based testing strategy to check the adequacy of the testing. The cost of quality can be extremely high, so a risk-based testing strategy can be used in strategic change and software activities to reduce time and cost while effectively removing critical defective outputs.

In theory, not all aspects of a solution goal would have the same effect if that aspect failed. The failure of some parts of a strategic change programme would have more significant negative impacts than others. This approach may go more expansive than simply monitoring key results to other aspects of the change programme. A risk-based approach seeks to maximize the impact of testing by evaluating the solution goal and deciding which parts should have the more substantial investment in evaluation or testing. Risk-based testing can be used to mitigate a situation or even to avoid undesirable circumstances.

The quality process employed in strategic change activities is contextual and driven by Outcomes or Key Results. So, for example, solutions that could endanger the future of the whole organisation will have different quality regimes to those activities designed to alter an organisational structure.

All solutions, whether strategic change, software, or a combination, require a business-driven test management approach involving results, risk, time, and costs. This approach should align with agile principles, especially incremental delivery and continuous integration.

Not all teams are the same. A collocated team comprised of highly skilled and experienced individuals will have a different quality planning need than a geographically dispersed team with less experienced players. Teams in a regulated environment will be pressured to illustrate their quality rigour, while those in other industries will not. Therefore, a Quality Plan is needed that balances the value delivered, the operational and reputational risks, and the delivery cost.

This Quality Plan outlines the tools, techniques, data and strategy for the testing. The key results and acceptance criteria will be broken down into a testing schedule with test cases linked to the Increment Roadmap and release plan. An agile quality plan employs a range of principles and techniques. Including:

- **Build-in quality** Agile is concerned about quality in fitness for use rather than conformance to some requirement specification. Built-in quality is defined in each work item's key results and acceptance criteria. The criterion are used when developing the output, and the Iteration Review provides the essential feedback element fulfilling agile quality management
- **Rapid feedback,** the short time box approach and the concept of demonstrating working tested outputs provide the quick feedback loops advocated by empirical control. While the measures and sensing used for OKRs provide strategic change activities with verification. Continuous integration and test automation techniques offer the same assurance for software activities.
- **Agile test strategies** Agile testing happens at multiple levels at the task level in all activities. In strategic change activities, outcomes are defined in horizon-based plans and themes. Business activities may also use impact mapping to help teams anticipate what might prevent them from achieving their goal and allow them to test accordingly. In software, quality is rolled up through testing of stories, features, iterations and finally, at the increment release.
- **Shift-left testing** the concept of testing an outcome, feature, or story as early as possible in the delivery process is known as shift-left. Work items that the team believes are finished but not thoroughly tested represent a delivery risk, as rework may be required. This rework would be unplanned and may impact the team's ability to complete activities in the subsequent iteration. Shift-left uses testing principles using personas, scenarios, jobs, or roles.

The team should be designed to allow shift left quality practices and solution evaluation using the key results within each iteration boundary.

Performance improvement

A fundamental principle in Lean is continuous performance improvement. Agile Lineout takes this theoretical stance and makes it practical by being explicit about context, inspection, and enhancement.

Unfortunately, all human systems are constrained, making perpetual improvement impossible. The Team Coach's quest is to identify correctly what constrains their team. Using data such as Lead Time and Cycle Time and tools like control charts and cumulative flow diagrams, the Team Coach identifies potential areas for performance improvement.

In changing the enhancement activity's focus, I have attempted to make the outputs measurable improvements. For example, how do we improve quality as a clear metric possibility? How we can improve effectiveness assumes that the team already has a measure they wish to change. Undertaking experiments using OKRs provides a metric that, when moved positively, illustrates performance enhancement.

The Team Coach challenges the team to improve performance faster by presenting data and facilitating discussions. The metrics used could be expressions of Cycle Time or Lead Time and for Scrum Team Velocity. The presentation of one or more trend analyses will often provoke serious discussions.

Team coaches should resist any attempt to stretch the team's delivery commitment except by experimenting with different ways of working. However, another output from the Enhancement Workshop, Context Assessment, or Inspection activities may result in new learning objectives to improve performance.

Coaching the Development Line focuses on effective product delivery. The emphasis is on building efficiently and using metrics to suggest measurable improvements.

Artefacts in the Development Line

The following table shows how Agile Lineout artefacts support the Development Line element. The table is intended to illustrate primary usage, but a team could choose any artefact to be used more extensively than I have suggested.

	MOBILISATION & PLANNING	INSPECTING PERFORMANCE	ENHANCEMENT WORKSHOP	INSPECTING COMPLETION	DELIVERY CONTEXT
SOLUTION BACKLOG	(✓)		(✓)	(✓)	(✓)
INCREMENT ROADMAP	✓	(✓)	✓	(✓)	✓
IMPROVEMENT BACKLOG	(✓)		✓		(✓)
ITERATION BACKLOG	✓	(✓)	✓	(✓)	✓
QUALITY PLAN	(✓)	(✓)	✓		(✓)
COACHING PLAN	✓		✓		✓

Figure 11 Artefacts used in the Development Line

Detailed descriptions of each Agile Lineout Artefact appear later in this book. The emboldened ticks show the strategy which creates or maintains the artefact. The shadowed ticks show artefacts considered or serve as inputs to the Development Line strategic step.

Team Behaviour

The team behaviour element of Agile Lineout focuses on how the team is operating. In an agile team, people must work together to perform their tasks rather than act as individuals. As shown in Figure 12, the Team behaviour line has four strategies; team building, team working, external relationships and learning. These team strategies are built on the foundations of; psychological safety, clarity over team roles, reducing silo behaviour and transparency.

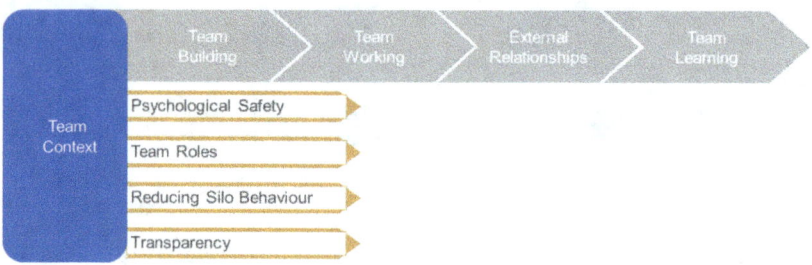

Figure 12 The Team Behaviour Line

At the outset of an initiative, a new team initially acts as a co-acting group instead of a team. A co-acting group may have the same purpose and proximity to one another, but they act like they each have a particular or individual job. Therefore, designing the work for a team rather than work for individuals increases synergy and performance and is vital for team building.

American psychologist Clayton Alderfer states that team boundaries must also be defined to establish team membership. Boundaries create the sense of being part of a team in the mind of each individual. They also demonstrate separation from traditional organisational structures and responsibilities.

One further membership consideration is team size. Robin Dunbar states that a group of up to fifteen people could be considered a team. The team size dictates the team design, roles and responsibilities, and ways of working considerations. When evaluating their team context, a team will consider their levels of collaboration, mutual support and the performance of individual actions within the agreed roles and responsibilities. In addition, they will identify needs for further team building, changes to how they are working, alterations or strengthening external relationships or changes to improve the team learning.

Team building

With newly constituted teams, team building becomes essential. Team building refers to activities that enhance social relations and define roles necessary to turn a group into a team. However, it cannot begin until the membership and structure of the team are established. Therefore, these aspects need to be clarified as early as possible. If unresolved, ambiguity regarding membership will always cause team behavioural issues as the activity progresses.

In a new-to-agile company, a team may find resistance to its self-managing behaviour and actions. Past working practices suitable for traditional delivery methods often inhibit agile practices. De-limiting a team by removing past ways of working is a frequent coaching challenge.

Sometimes, the roles supporting traditional approaches are no longer necessary. These conventional roles will need to be explicitly altered by the application of new governance mechanisms and responsibility models.

Alternatively, some supporting roles such as specialists, lawyers, accountants, architects or domain experts may be invaluable and essential to the team's work. However, these specialists may not be part of the core team and should not be included in Team Building. Nevertheless, they need to be identified as team members who need to know who they are and who to if they encounter difficulties.

Working as a team

Once the membership is established, team building often starts with understanding the roles and ways of working and documenting these in a Working Agreement or charter. These conversations allow a newly formed team to get to know each other and start developing relationships.

Sometimes, novice team members try to use the same work patterns as they did in traditional teams. Unfortunately, this behaviour results in silos and delivery inefficiencies and often results in an unhealthy them-and-us situation between team members, which destroys teamwork.

Therefore, the Working Agreement should document the Ways of Working defined in the Development Line. It may also note decisions about working together, such as meeting times, roles, and responsibilities. If the team is across multiple time zones, guidance may also be needed about core working times. In addition, the Working Agreement may outline the rules that they have agreed about the giving and receiving of feedback.

The Working Agreement should also outline the team design, roles and responsibilities. The team design element is essential when a team is part of an agile programme or big team. In strategic change activities, assigning a duty to monitor Risk is necessary. Finally, the Working Agreement may contain anything the team feels will help them work together.

External relationships

Agile is not a single practice; it is a frame of mind, principles, and values requiring individuals to become adaptive, learning and accountable. This behavioural change impacts team members and the people who surround them.

For example, some stakeholders with a traditional project approach may require a team to report and explain their decisions, actions, and performance. In addition, stakeholders who previously were comforted by the sight of a traditional project plan often need reassurance that an agile team has an equally disciplined process that merely produces outcomes differently. Therefore, the agile team must consider the needs of their stakeholders and act in ways that reassure and provide the information they need.

An agile team must also engage those stakeholders who must provide input and guidance on the business outcome. Aspects of the agile process critically depend on regular feedback from such stakeholders. These people must understand their role in providing input regarding the solution's suitability.

An agile team needs to understand how it fits into the broader organisation. It must constantly monitor the appropriate stakeholders' engagement, information needs, and attitudes. Finally, it should identify and manage all of its external relationships proactively.

Team learning

The phrase "fail fast and learn" is common in agile circles; however, little guidance is provided regarding how or when a team learns. Richard Hackman states, *"A team whose members learn together how to work together, and who then stay together to build their collective competencies, is almost certain to develop into a more effective performing unit than otherwise would be the case"*. Team learning is an essential element of Agile Lineout, with ingredients, artefacts and events designed to enhance learning and improvement.

In Agile Lineout, team learning items are placed on a learning parking lot at any time to be considered or recognized in the future. When using the parking lot, the team logs items to confirm things they have learned or identify training needs. In this way, the team will be reminded of their developmental progress and areas where they would like to know more.

Learning items logged in this way may not involve training. Instead, learning may happen using experimentation or the team simply taking time to explore an issue together.

The brief review of the team's learning at the start of each Lineout draws the team back to an agile team's behavioural aspects and ensures that the right foundations are in place.

Behavioural foundations

While the above team behaviour strategic steps present a virtual circle of opportunities to improve teamwork and learning, they are built upon foundations that need continual focus. The four foundations are:

- Psychological safety.
- Responsibility and accountability.
- Reducing silo behaviour.
- Transparency and Communications

Psychological safety

Psychological safety is perhaps one of the more difficult agile mindset elements to coach, yet it is a critical building block of the continuous improvement process. According to behavioural scientist Amy Edmundson, psychological safety believes *individuals will not be punished or humiliated for speaking up with ideas, questions, concerns, or mistakes.* It is also a shared belief among team members that others will not embarrass, reject, or punish them for speaking up.

If an organisation has a blame culture or shoots the messenger, Psychological Safety will take a long time to arrive! In other words, team behaviour requires psychologically safe partnering between the team members. In addition, for any conversation to yield successful results, people must feel safe enough to give feedback to say what is on their minds.

Psychological safety is an essential element of any agile team. Therefore, teams must guard against individual or group behaviour threatening this foundational agile principle.

Team Roles

In all teamwork, even with self-managing teams, team members need to understand and hold each other to account for individual responsibilities.

Agile teams are, by design, flexible and responsive. Team designs may involve distributed membership, travellers, ambassadors, or multiple squads. In these circumstances, roles and responsibilities tend to expand beyond those in the Scrum Guide.

At the most superficial level, stakeholders remote to the team may require summary financial and status reports. In addition, the team may need to design communications for the broader business or engage in some change management activities. Therefore, consideration of all the roles and responsibilities is required to assess the team context.

Beware of making someone responsible for something they are not personally involved in, as this generally leads to conflicts, quality issues and delays.

These roles and responsibilities do not need to be permanent or fixed. They may be adjusted at any time. However, documenting roles and responsibilities in the team working agreement is often necessary and may be a compliance requirement in some regulated industries.

As the activity progresses, the team may discover that new roles are required, or an individual may leave or wish for a role change. So, responsibilities and accountabilities may alter over time. If so, changes will be needed to the Working Agreement.

Reducing silo behaviour

Silo behaviour describes when individual team members draw boundaries around their roles and responsibilities, which artificially preclude them from doing other things. Silo behaviour is the opposite of team working, where backup or helping each other out is desirable.

While a Team Coach may be focused on reducing the impact of organisational silos, a newly formed team may silo team members based on their specialization or the function they previously performed. These team silos must be addressed as they represent a challenge to flow and efficiency due to waiting time and handoffs.

Silos can be reduced by cross-team training and peer learning. Although, at the same time, it is not always possible to quickly create T-shaped individuals with cross-functional skill sets, so pragmatically some inside team specialisations may be needed. Defining the agile ways of working presents an opportunity to reduce inside team silos and offer prospects for individual or team learning and development.

A team silo may be necessary as the organisation or team adopts new tooling or working methods. However, a team should guard against silos as a foundational principle to effective teamwork and effectiveness.

Transparency and communications

In newly formed teams, organisational or behavioural baggage often accompanies people. Some team members may initially find agile's openness and honesty elements difficult. This fear will be magnified if the organisation has a "shoot-the-messenger" culture. However, transparency is an essential modern management value needing a coach to address concerns supporting both the team offering the information and the stakeholders or recipients of the communication.

Earlier, I mentioned stakeholders in terms of feedback in the solution line. Stakeholders also have a part to play in coaching Team Behaviour because they have communication and change management needs. In this context, stakeholders are those individuals or functions that are impacted or need to prepare somehow. The coach must know that Stakeholder Engagement is essential to successful change, which is ignored at peril in some contexts!

The Solution Owner and team members must ensure regular, appropriate, and informative communications are made in many situations. Remembering that communication is a two-way street means that some teams will need to establish ways of listening in addition to those for speaking.

Artefacts in team behaviour

The following table shows how Agile Lineout artefacts support the Team Behaviour element. The table is intended to illustrate primary usage, but a team could choose any artefact to be used more extensively than I have suggested.

Detailed descriptions of each Agile Lineout Artefact appear later in this book. The emboldened ticks show the strategy which creates or maintains the artefact. The shadowed ticks show artefacts that are considered or serve as inputs to the Team Behaviour Line.

Some coaches may find documenting aspects of team building a little strange. However, I have found it invaluable, especially when there is a need to change a Team Coach or when a team has new joiners or leavers.

I have also found these documents invaluable when team members are not working in their native language. For example, sometimes team members need to translate words or documents so that they may fully understand.

Writing information and dates on the evidence board reminds a team of what they have achieved. Likewise, the Learning Parking Lot reminds the team what they wish to learn or improve.

So, while I would support brevity, I have found the effort to document team-building information on a radiator or in a tool like Mural invaluable.

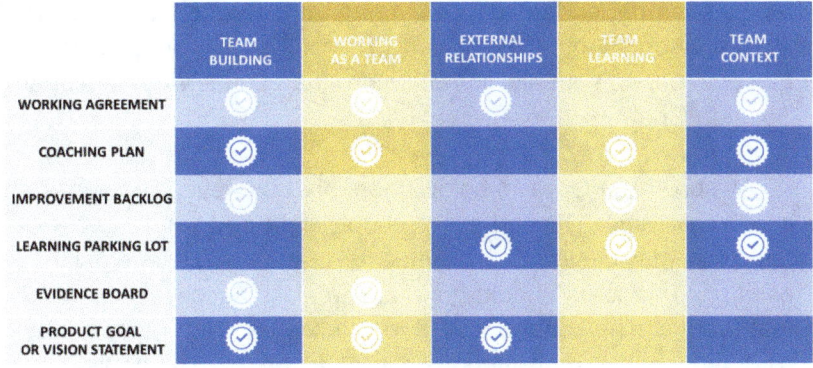

Figure 13 Artefacts in Team Behaviour

Scaling Agile Lineout

Sometimes an initiative will require a big team or programme. As team size grows, splitting or segmenting the team into a collection of smaller teams organised by functions, specialities, or workstreams is common. Deciding how to separate the teams is critical in a programme, as a poor design will result in dependency management and coordination challenges.

In Agile Lineout, I have suggested an optimum team size of fifteen, based on Robin Dunbar's research. However, the need for splitting a team often depends more on activity context rather than the number of people. Logical splits may be used as an alternative. For example, an acquisition could be made up of three teams. The first is the people managing the HR aspects, including payroll. The second may be people concerned with financial integration. Lastly, there could be a team undertaking IT due diligence. Three teams in the acquisition programme may be needed if these activities have significant overlaps or dependencies.

The first rule about creating agile programmes is don't! The ideal scenario is to design the work so that overlaps are removed and teams are independent. If, however, more than one team is needed, Agile Lineout should be scaled. Therefore, this section of Agile Lineout concerns an activity with more than one team with a shared Solution Goal.

As shown in Figure 14 below, there are two ways to design teams in a programme. The first and probably preferred way is to create a team that may design and deliver a full feature or discrete part of the solution. This team should have all the skills necessary to develop that feature. The second way is to establish a team responsible for a component that may support several features. An example may be an HR element involved in several parts of an acquisition or organisational change activity. Again this team should have people with all the skills required to support each feature from a component perspective.

In both instances, a team is cross-functional, potentially comprised of individuals from various line-management silos. However, the principle behind self-managing teams is to give them

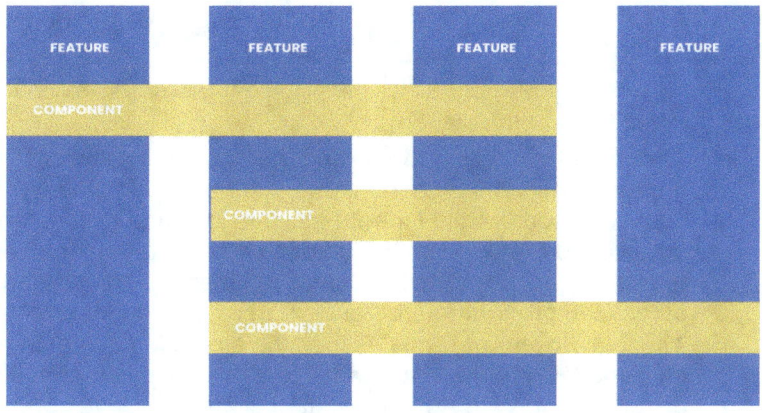

Figure 14 Constructing a Programme

managerial responsibilities. Therefore, the programme design should keep the organisation as flat as possible to empower the teams. Sometimes a programme may need a hybrid design combining teams to deliver features and components.

Each requirement area should have a Solution Owner responsible for identifying work items for their teams and aligning them with other solution owners' outputs.

The essential element of an agile programme is coordination between the teams. Coordination happens with the three lines of Agile Lineout, the solution, the development process and the teamwork.

Due to the coordination requirement, an agile programme needs an additional role, the agile programme manager. An Agile Programme Manager is a leader, sometimes a servant, responsible for optimising the strategic change delivery and maximising the organisational value created.

Scaling the Agile Lineout Framework

The Agile Lineout principles used for multiple teams in a programme are identical to those used with a single team. The only difference is the changes needed to coordinate and synchronise the delivery system using several teams. However, we still have the three element lines of the system:

- **The Solution Line** Identifying strategies to create the appropriate high-value outcome and the foundations for continual focus. Agile Lineout at scale includes actions and activities to break down the solution for delivery by multiple teams. It also provides the review and feedback on artefacts across the programme rather than at a single team level.
- **The Development Line** Outlines the strategies to create an efficient delivery system with its continual focus on effective development. With Agile Lineout at scale, the Development Line includes coordinating working methods and planning across all teams to optimise the delivery system.
- Lastly, the **Team Behaviour Line** enables both solution creation and development efficiency. The same is true with Agile Lineout at scale. Working with bigger teams or teams of teams emphasizes enhancing teamwork and establishing programme-level roles and responsibilities with programme-level Working Agreements.

Hence, in Agile Lineout at scale, we still have the Solution Line and the Development Line, enabled by Team Behaviour. In addition, as shown in Figure 15, extra foundational elements are added in the form of:

- **Coordinated Planning** Concerns creating plans which are synchronised and consider dependencies both from inside and outside of the programme
- **Activity Control** Relates to the additional activities of Inspecting Performance and Inspecting Completion at the programme level
- **Continual Integration** Concerns merging and ensuring the quality of programme outputs produced by the interdependent teams. It considers the continual merger of the team creations so they may be released together frequently.
- **Frequency of release** As a programme is a more extensive delivery system; there may be a temptation to slow down the flow of value. To avoid this propensity,

programme deliveries should be small, low-risk, and incremental. Deliveries should be made as early as possible, providing benefits or capabilities as soon as practically possible to the organisation

The Agile Lineout at Scale, therefore, uses the same delivery system.

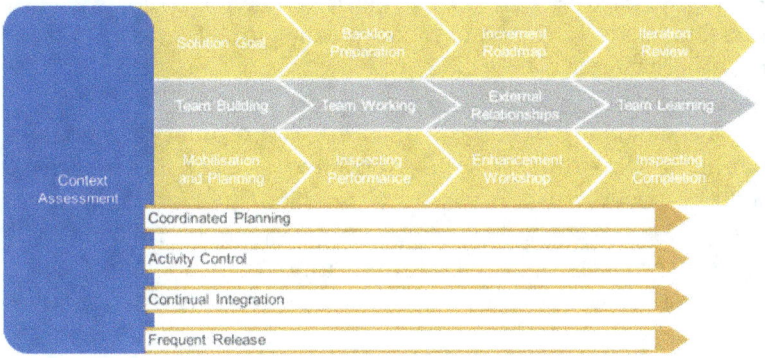

Figure 15 Scaling Agile Lineout

Scaling the Context Assessment

In Agile Lineout at scale, the Context Assessment is used at the programme level, with the three contexts being viewed for the entire initiative.

In the Context Assessment, a programme evaluates the situation and adjusts plans accordingly. The Context Assessment is linked to the Enhancement Workshop as a mechanism for continuous improvement. In this instance, improvements are being sought at the programme and team levels.

The Context Assessment may be treated as a separate activity or part of the introduction to Multi-team Planning.

Scaling the Solution Line

As stated previously, a Lead Solution Owner may be needed in an agile programme. However, this leadership could be provided as a function rather than a role. For example, in business transformation activities, a group of stakeholders or the Guiding Coalition for the initiative may undertake the part of the Lead Solution Owner.

In addition, the Backlog, the Increment Roadmap, and the Iteration Review must be aligned, synchronised, and coordinated across several teams. For example, the team Solution Backlogs must be aligned into one or more releases to deliver minimum viable increments.

The Solution Owners of each sub-team must coordinate their activities to ensure that the developed solution is complete and without waste. For example, the Solution Backlog refinement will involve teams selecting work items or dividing the Backlog by team makeup or skills. In addition, the Programme Solution Owner will need to ensure that dependencies and handoffs can be managed as work items are selected.

Joint Solution Design uses the joint application design (JAD) technique. JAD is a process from the dynamic systems development method (DSDM). It is a technique used to collect business requirements with several teams of people. It includes approaches for enhancing participation, expediting development, and improving quality. It consists of a workshop where teams meet, sometimes for several days, to define and review the business requirements. The Joint Solution Design approach enables collaboration and encourages teams to work more effectively and shorten the requirements definition time frame

The Iteration Review happens at the programme level, with feedback distributed to the relevant teams. Depending on the context of the initiative, these teams may or may not hold individual team Iteration Review sessions. In line with good agile practice, all the solutions produced in the Iteration should be integrated and tested before the Programme Iteration Review.

Scaling the Development Line

In coordinating the programme delivery processes, teams must agree on ways of working, the use of tools, communications practices, and essential record keeping. Mobilisation and planning will synchronise the delivery cadence, release plan, and associated quality and integration activities.

Multi-Team Planning is a straightforward yet compelling technique. Multi-team planning is what it says. All the teams and anyone else needed to collaborate virtually or physically, to coordinate the high-level plans. The planning activity develops a shared understanding of how they'll achieve outlined business goals over a planning horizon. Of course, planning horizons vary, but in both software and non-software initiatives, teams usually plan their work for the next quarter of a year.

Multi-team planning is expensive due to the number of people involved, so some organisations plan quarterly. However, a prerequisite for quarterly planning is a reasonably stable backlog. A team with a backlog that changes frequently may need to plan for shorter periods.

A requirement for recurrent planning may necessitate planning by representatives rather than a whole team approach. However, there are substantial psychological benefits in team members creating high-level programme work plans in addition to the specific detailed plans of their team.

The outcome of Multi-team Planning is an Increment Roadmap showing deliveries based upon aligning the workload to the team's capacity and considering internal and external dependencies. The goal is to align the activities in each team to create viable increments.

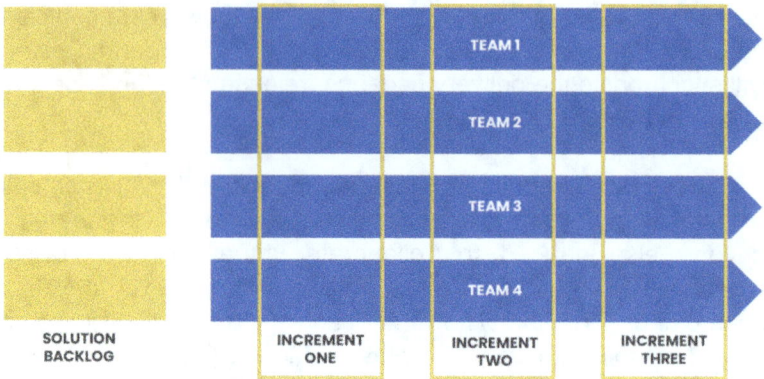

Figure 16 Increment Alignment

The output of Multi-team planning is a single roadmap of incremental deliveries harmonised across all of the teams. See Figure 16 for an illustration.

Inspecting Performance is a question of each team considering working in a daily inspection activity with a less frequent but essential inspection at the programme level.

Programme teams tend to use a Programme Kanban. This Kanban Board aggregates data from the individual teams providing an overview of the programme delivery system. It will major in showing dependencies and the Increment Roadmap.

An Agile Programme team may place WIP limits on the programme board and use a cumulative flow diagram to identify programme-level constraints and delays.

When delays or issues are encountered, the programme team may use action teams or swarming to resolve challenges.

Scaling Team Behaviour

Scaling Team Behaviour is primarily a question of defining the additional roles needed by a programme and developing a culture of collaboration.

For single teams, Team behaviour has four strategies. Team building, team working, external relationships and team learning. The team strategies are built on; psychological safety, clarity over team roles, reducing silo behaviour and transparency. With an agile programme, nothing changes except magnitude.

Team Building will be needed at the programme level in addition to the activities within each team. Programme-level decisions will need to be defined in Working Agreements. The importance of External Relationships increases with significance in line with the size of the undertaking. Each programme also needs to learn as a team, which is often more challenging to engineer due to scale.

Scaling Agile Lineout foundations

While I have outlined the similarities with each line and alterations to the strategic steps, I have identified that a programme requires additional focus areas. In my experience, running a successful programme requires four foundations.

- **Coordinated Planning** A programme must continually assess if its coordinated plan fits its purpose. In strategic change activities, situations may be encountered, expectations may alter, or stakeholder feedback may cause a team to pivot. The consequence is that although the programme undertook multi-team planning, that plan may have lost its currency. The Inspecting Performance and Inspecting Completion activities at the programme level provide focus and analysis by continually asking if the plan will enable the endeavour to reach its Solution Goal.
- **Activity Control** With a larger team, activity control becomes very important. The probability is that there are more dependencies to manage, calling for diligence from the Agile Programme Manager to spot adverse trends and seek resolution to potential problems before programme performance is impacted.
- **Continual Integration** If several teams are involved in completing a solution, then there is a greater risk that the items delivered will not fit together. The longer work items are created but not assembled and tested, the

greater the risk. Since Agile Programme Management is about reducing risk to increase the probability of success, continual integration is of great concern. So alongside the Increment Roadmap, an Integration and Test Plan is an essential artefact for a programme team.

- **Frequent Release** Unlike a software project, a strategic change team may need to frequently place programme deliverables in the hands of customers or stakeholders. For example, a cultural change programme may deliver communications followed by training and an expectation that people will make a change. Change readiness becomes a factor, as does change fatigue. In other words, a business unit may need to be assessed if it is able or appropriate for that unit to make a change. Often situations outside of the remit of the agile programme may cause a programme team to rethink. I am reminded of a change team I was working with where we were ready to implement some new accounting software. However, several senior managers had recently resigned from the business unit, which was effectively leaderless. Clearly, changing that function's critical business software was inappropriate until leadership was re-established in that unit. Our Change Readiness assessment established these circumstances and recommended a delay to the project team.

This section should be read in conjunction with descriptions of each delivery line earlier in this book.

The Lineout Artefacts

As previously indicated, Agile Lineout focuses on creating team effectiveness. This list of artefacts is designed to help the team quickly and efficiently produce high-quality outcomes. In addition to the artefacts listed below, the team will use a task board or Kanban board to visualize its work. Large teams will also use a programme board. This list is a minimum set; additional artefacts may be required in certain circumstances.

Working agreement

A Working Agreement is a document that lists a team's goals, behavioural norms, accountabilities, responsibilities, and working practices. For example, if team members are distributed in different time zones, the Working Agreement may record the core working hours arrangements. Core working hour agreements set parameter times before and after which team activities are inappropriate.

A Working Agreement removes assumptions and makes them explicit so the team can remain aligned throughout the initiative. A Working Agreement can be a handy tool when team building as a new team is spinning up and "norming" together or if conflict situations appear. The debates and agreements needed to create the Working Agreement allow a new team to get to know each other and resolve some potential conflicts.

Agile Lineout takes a Working Agreement a stage further to get the team to think through their goal. What type of team do they wish to become? What reputation would they like? What reputation do they want? For what would they like to be known? Establishing a team goal is the first step in the *GROW model* created by Sir John Whitmore. After that, the team uses the remaining R reality, O options, and W what will you do steps during the Enhancement Workshop.

Coaching plan

A coaching plan gives structure to the team development activities. It starts from a generic "first meeting" situation and then adapts as the activity progresses and the team learns. The coaching plan is modified regularly by the team. This plan identifies the need for coaching and training over time. Initially, this plan will be generic. However, with each iteration, the Team Coach's plan will become tailored to the needs of a specific team and the situations they face. Over time, the coaching plan will evolve based on the team's improvement options identified in their Enhancement Workshops.

Improvement backlog

The Improvement Backlog is a list of options the team has selected as either an experiment they wish to try in the future or improvements they want to make. Items on the Improvement Backlog may come from the Enhancement Workshop, the team Learning Parking Lot (see below), or from improvements the organisation may be introducing to the agile ecosystem, such as implementing a new process or a tool.

The team may select items from their Improvement Backlog during their Context Assessment and plan them into an Iteration during Mobilisation and Planning.

Learning parking lot

The Learning Parking Lot is a place to capture comments, topics, questions, and items that the team has learned or wishes to learn. While working, the team players may discover or learn things during daily interactions or iterations with stakeholders or those external to the team. Alternatively, these interactions may reveal an opportunity for a team to learn something new. Either way, these discoveries may be recorded on the Learning Parking Lot and considered by the team to assess their learning context at the beginning of the next Iteration.

Evidence board

Detectives use an evidence board to evaluate a crime. An agile team uses it to illustrate achievements and remind them where they started. A quotation attributed to Henry Ford says, *"Whether you think you can or can't, you're right."* It's about positive thinking. An evidence board reminds teams of their journey and the lessons they have learned. It provides a trail pointing towards the team's North Star, its reputational goal. The Evidence Board may also form part of the team's agile radiator.

An evidence board can be physical or virtual, perhaps using Mural or Miro, which displays images, graphs, emails, slogans, or anything the team decides is appropriate. Each piece of evidence must be dated. When the team questions its attainments, which they often will, they can refer to the evidence board as a source of inspiration and focus.

Dating the evidence can reveal patterns or gaps in regular achievements. The dates may show, for example, that the last quarter had a lot or few accomplishments. The evidence board is a means of team motivation and serves as a motivator and a reminder.

Solution goal

The Solution Goal is a statement that describes the overarching objective or vision for the solution, service or strategic change activity. Some vision statements are aspirational, while others, such as regulatory compliance, are factual. The solution goal lays out what the organisation hopes to achieve as the outcome of the agile activity.

Increment Roadmap

The Increment Goal represents the significant accomplishments needed to make the Vision a reality. The chain of Incremental Goals describes the Increment Roadmap highlighting the steps by which the outcome will be delivered and will support or add value to the organisation.

The Increment Goal should be easy to understand, and in some instances, it may be necessary to describe the goal by the features it may contain or the outcomes it is planned to achieve.

Solution backlog

The Solution Backlog lists the work items needed to deliver the solution goal, the outcomes and key results. It may also contain tasks such as communications, workshops, specific testing activities, change readiness assessments, and organisational or infrastructure changes. A critical agile principle is that team members only work on the Solution Backlog items. So if work is needed to be completed by the team members, it should be on the backlog. The reasons for this strict rule are that the team needs to understand its actual capacity and give all team members and stakeholders visibility of the totality of the work being completed.

The Solution Backlog is a prioritized list. The prioritisation may be based on the value of the work or risk reduction. The Solution Owner and team will prioritise items on the Solution Backlog until the budget or time available for the product has expired or the cost of completing any additional work item exceeds the anticipated value created. This means it may be decided that some things on the backlog are not completed because they are of insufficient importance,

Iteration backlog

The Iteration Backlog lists tasks identified by the team to be completed during the next iteration.

During the mobilization and planning meeting, the team selects features or OKRs from the Backlog and breaks them into work sized to fit within an iteration. In software teams, features are usually broken down into user stories. However, in strategic change initiatives, the backlog is the list of transformational activities resulting from analysing the OKRs and deciding the work required to achieve the desired result.

Quality plan

Like software projects, strategic change programmes also require a Quality Plan. The quality assurance activities in strategic change ensure that the team outputs have had the desired result. Consequently, the plan is contextual and business-driven. The plan identifies.

- the expected results,
- the risks which are being assessed,
- the timing or nature of the tests, and,
- the tools or techniques to be employed.

The Quality Plan must be adaptive and evolve as the activity progresses and situations are encountered. In specific regulated industries, a Quality Plan may be scrutinized for suitability and its level of rigour.

It is easy to think that testing only applies to software teams. However, if a transformational team decides to alter a process or function, it must also ensure its change has the desired result. Software activities have the concept of regression testing. In strategic change activities, it is also necessary to check that the subsequent alteration made by any team in a programme has not had a detrimental effect on a previously verified outcome produced by any team.

In the world of strategic change, we also have the human factor. When individuals or previously altered functions are stressed, their behaviour often reverts to previous patterns. This means a strategic change programme must constantly monitor, measure, or obtain feedback that the results our outcomes previously assessed as being completed have not been altered or regressed.

Lineout Metrics

As variously previously stated, the team is an organisational construct, but it is also a system. The team takes features or outcomes their stakeholders request and converts them into reality, creating value as it does so. The team's actions are a system that can be managed, developed, automated in parts and optimized. In principle, every system is bounded by space and time and constrained or influenced by its environment. A system is defined by its structure or purpose and expressed through its components, functions and performance measures.

Agile Lineout uses metrics to plan and check the progression towards the desired outcome. Most metrics should be actionable, meaning the team does something based on the presented data. Actionable metrics are distinct from vanity metrics which typically make someone feel good but do not enable progress towards a goal or indicate what to do next.

In addition to the metrics identified below, the team could use burndown or burnup charts, cumulative frequency charts, dependency diagrams, and control charts. The metrics used by a team could be captured as part of regular business reporting. Or may necessitate the team to create reporting capabilities to evaluate the impact they are having.

Value created

Value creation is a crucial concept in agile. Identifying and focusing on value enables the team to select the most critical activities to generate the most benefit and move the team closer to the Increment Goal.

Sometimes the use of an absolute number, such as monetary value, is inappropriate or irrelevant to the activity. Instead, a Solution Owner may use a synthetic measure such as value points. Value points can be created using the same techniques a Scrum team may use to estimate story points.

In conjunction with the team, the Solution Owner uses value to prioritise features in the Backlog and uses dependencies and risk to determine the sequence of work completed. Common sense suggests that any activity deemed high risk by the team should be planned as early as possible in the delivery cycle to avoid waste should that risk occur!

Progression metrics

Progression metrics or key results (from OKRs) are data provided to the team to evaluate the impact of their last increment. For example, a team working on an initiative to increase customer satisfaction may monitor customer net promoter score (NPS). If they release an Increment and NPS improves, they know they are progressing towards their desired outcome. Alternatively, if the NPS score fails to change or is reduced, it may cause the team to pivot or alter its approach.

Progression metrics are not internal to the team but are typically generated in the market or by the organisation's business activities. These metrics show how the team influences an outcome but could also show how other factors influence it. For example, progression metrics could show the impact of a new competitor joining the market. In strategic change activities, the OKRs will show if the team's actions have had the desired outcome. Evaluation of the metrics may indicate a need to alter the solution goal, change the Backlog, or alter the Increment Roadmap. Progression Metrics are also used during Mobilisation and Planning.

Lead time

Lead time is defined as the time between the start of a process and its conclusion. For example, the feature lead time is defined as the time between work starting and when that feature begins to deliver value to a customer. A lead time comprises the cycle time for each step in the delivery process and the waiting time for each step to occur.

Monitoring Lead time in an agile team is essential as it shows how quickly value is created. The size of work items negatively impacts lead time. In theory, the smaller the work item, the faster it is produced. An analysis of lead time could show issues with the solution design or the delivery process. Using lead time without other metrics to put the measure in context can be misleading. For example, lead time may be used as a primary measure of team effectiveness when combined with value created and displayed on control charts.

Number of items in process

In Kanban, work-in-process (WIP) limits restrict the maximum amount of work items in the team's value stream stages. A WIP limit constrains the number of work items a team may work on simultaneously. They enable flow by managing capacity on team Kanban boards.

Using work-in-process limits allows the team to complete single work items faster, focusing only on finishing tasks rather than starting something new. In addition, work-in-process limits enable the team to identify constraints or bottlenecks in their delivery processes.

WIP limits are difficult to implement because limiting work in progress seems counter-intuitive to traditional managers. "Do less to get more work done?" Yet, the truth embodied in lean is immutable. Actively manage how many work items a team of people is working on at any given time, at both the individual and team levels, then a focus can be created to get work done quickly, with high quality. Lean means that they get more done overall by doing less work simultaneously.

A WIP limit forces a team to decide the priority, time sensitivity, and cost of delay of various work items. If a team never hits its work-in-process limit, it is probably a sign that the limit is too high or that too many people are waiting for work.

Flow efficiency

The Flow Efficiency metric examines the two essential components of lead time: working and waiting times. The key to efficiency is to calculate how long work is spent waiting. If the team uses a Kanban or task board, it will be structured or configured to show when a work item is queued and ready to be pulled to the next step and completed. Therefore, measuring how long work spends in those queue columns is needed to calculate flow efficiency correctly.

Quality metrics

Strategic change activities are frequently asked to measure their efforts' results and report back on them to stakeholders. Yet these measures are also used by the strategic change team to evaluate progress or alter direction. Strategic change measurements usually include surveys, tests, assessments, and completed activity observations. It's also essential to track the success of business outcomes.

Quality metrics are crucial for agile teams. They help a team visualise how their delivery system has performed over a specified period. That period could be whatever the team wants: a single day, a week, a sprint, a release, a quarter, or an entire activity. Furthermore, the methods used to measure the impact usually combine quantitative and qualitative metrics. Each technique offers a different perspective on the results.

- Quantitative measurements count and records factors such as:
  o the average time it takes for people in the company to adopt the changes
  o the number of people who actively apply the changes in their work
  o the level of success employees achieve when they implement the changes
  o whether the change management strategy is on track to achieve the desired outcomes.
- Qualitative measurements are more subjective and harder to measure. It usually involves methods like;
  o surveys,
  o performance reviews, and
  o quality assessments.
- To deliver strategic change, teams need to find out answers to questions such as:

- How comfortable do people feel about the strategic change?
- How satisfied are they with the communications they received?
- Do they believe their training in the new methods was enough to allow them to do their work correctly?

By measuring the effectiveness of the strategic change activity, a team establishes its efforts' success rate, whether it has achieved its goals and whether improvements are needed. Therefore, quality metrics are considered in the Enhancement Workshop by enabling the team to evaluate the overall effectiveness of their delivery system.

Capacity

Calculating capacity focuses on determining how much work a team can complete. The objective is to maximise the value created in each iteration by aligning workload to capacity. Too much work and the team will be overloaded, context switching will occur, and the Flow of work finished will be slowed. Conversely, too little work and team members' time will be wasted, and capacity will be reduced.

In software teams, capacity can be established by using velocity. In strategic change, capacity may not be so rigorously determined, and WIP limits can be used to control the workload. The principle of pull and average activity completion time can be used in Kanban to calculate capacity. Establishing capacity is essential for planning and monitoring the progress of strategic change activities.

How metrics support the elements of Agile Lineout

The following figure shows how the above metrics support elements of Agile Lineout.

	PRODUCT GOAL OR VISION	INCREMENT GOAL	INSPECTING PERFORMANCE	ENHANCEMENT WORKSHOP	INSPECTING COMPLETION
VALUE CREATED	✓				✓
PROGRESSION METRICS	✓	✓			
LEAD TIME			✓	✓	
ITEMS IN PROCESS		✓	✓	✓	✓
FLOW EFFICIENCY			✓		
QUALITY METRICS			✓	✓	✓
CAPACITY		✓	✓		

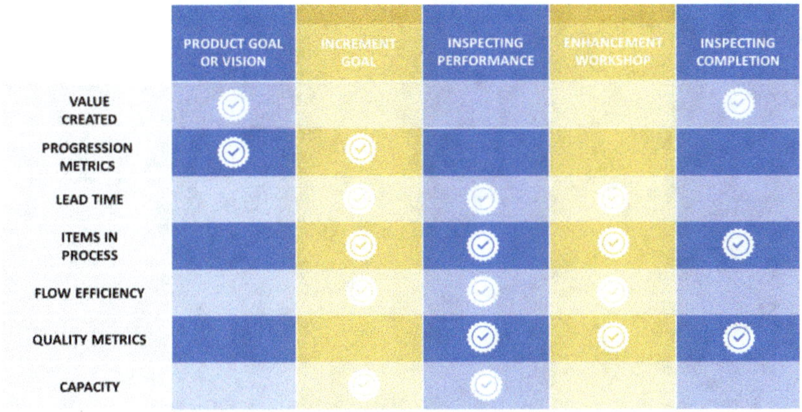

Figure 17 Metrics used in Agile Lineout

When defining the solution goal, the Solution Owner should be clear about the value expected to be created. The OKRs or Progression Metrics are used to illustrate progress towards the goal.

The Increment Goal is supported by five of the metrics. The deployment of the increment could alter the Progression Metrics. The team can monitor the Lead Time, so they may analyse their performance and seek improvements.

Items in Process and Flow Efficiency indicate their delivery process's anticipated and actual performance. Capacity is initially used in Mobilisation and Planning and has implications for the Increment Roadmap.

Inspecting Performance is the daily activity in the first half of the iteration, where the team decides if their performance requires adjusting the iteration plan. In making decisions, the team uses several metrics to give a picture.

The Enhancement Workshop may use any metrics to decide how the team may improve and what changes they need to make to reach their goals.

Inspecting Completion is the daily activity in the second half of the iteration, where the team asks itself, "will we get there?" Again, the team uses a range of metrics and things like burndown or up charts to illustrate their progress.

Coaching Agile Lineout

This section has been written for coaches who wish to implement Agile Lineout. It is not part of the Agile Lineout framework but rather a guide to implementation and getting started.

I have extensively used Dr W. Edwards Deming's System of Profound Knowledge (SoPK) in creating Agile Lineout. Although I do not suggest that a coach explicitly teaches SoPK to a team, I think it can be used to define, analyse, and create coaching plans and team performance.

Profound knowledge is a management philosophy grounded in systems theory. SoPK is the culmination of Deming's lifelong work. It provides a framework of thought and action for any leader wishing to transform and create a thriving organisation. The theory asserts that every system has four parts:

- **Understanding the system** Every development activity is a system where an action in one part will affect the other parts. Sometimes these are called unintended consequences. Deming asserts that individuals can better avoid these unintended consequences and optimise the whole system by learning about systems. In Agile Lineout, I have made explicit the elements of the delivery system and explained how these elements might interact.
- **Knowledge of variation** When teams create a hypothesis, a plan, or a statement of the expected outcome and this expectation is not met, Deming calls this a variation. Individuals and teams who do not understand variation inadvertently increase it through actions. "Why was our expectation missed? Why did something go wrong?" "How can we repeat this success?" The knowledge of variation results from asking these questions, finding correct answers, and taking the right course of action. The Inspection Activities are designed for the team to ask these questions. Analysis of a delivery system shows two types of variation: common cause and special cause. Common cause variations are the natural result of the delivery process and will be predictable within certain limits in a stable system. Special cause variations represent a unique event outside the system: for example, a natural disaster. Focus on common cause

variation through team coaching enables refinement and optimisation of the delivery system over time.

- **The Theory of Knowledge** Deming suggests that there is no knowledge without theory. It is based on understanding how people think or act and what they believe to be true. Beliefs are core to the theory of knowledge. Hence, asking questions such as; How can we avoid the mistakes we are in danger of making? How can we improve the learning process? Encourages a team to learn. Deming advocates experiments, prediction, and learning. In the Enhancement Workshop, a coach should encourage measurable improvements and use metrics to validate the actual outcome.

- **Understanding of Psychology** To understand the interaction between work systems and people, leaders must seek to answer questions such as: How do people learn? How do people relate to change? What motivates people? This understanding creates the knowledge that people are not cogs in a machine and that every team member brings extraordinary talents and abilities to the organisation. It is incumbent on every coach to ensure that these assets are not wasted.

The SoPK is a way of looking at Agile Lineout and how the team operates when considering performance improvement and always remembering that unintended consequences also provide a learning opportunity.

Coaching the Context Assessment

The Context Assessment uses the feedback theory of Systems Thinking and quickly checks all aspects of the delivery system before embarking upon the next iteration.

The Context Assessment allows single-loop learning using Deming's PDSA cycle. The PDSA cycle involves changing methods and improving efficiency to obtain established objectives (i.e., "doing things right"). It also provides an opportunity for Double-loop learning. Chris Argyris created this concept in the mid-1980s, which is different from single-loop understanding.

Double-loop learning concerns changing objectives (i.e., "doing the right things"). During the Context Assessment, a team may ask, "Why are we doing this?" and then decide to pivot if no or little value is identified.

In many instances, I have found that the context of a strategic change team alters after they deliver an increment. A change team will wish to measure the effects of its activity and evaluate if the planned outcome has been reached. This team will then adjust its plan based on their observations and feedback.

Coaches using Agile Lineout have suggested that the Context Assessment is a valuable second short retrospection before kicking off the next iteration. I agree with this view and think it mirrors what Scrum teams do before sprint planning.

If an existing team wishes to adopt Agile Lineout, the starting point is a context assessment emphasising the solution goal and then the Team Behaviour Line. See Figure 18.

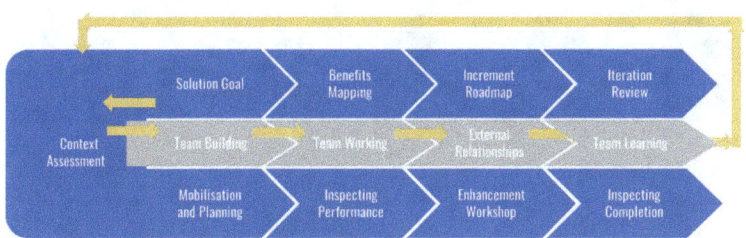

Figure 18 Starting Agile Lineout on an existing assignment.

Are the people functioning as a team? For example, have they established appropriate roles and responsibilities, and a good working agreement exists? How are they working with people outside of the team, for example, stakeholders and other groups in the organisation where they need interaction? Then, do they have the right skills to deliver the Solution Goal

The sooner a team creates a Learning Parking lot, the better for a new coach. The Learning Parking lot shows what team members consider as their development needs. Learning needs may be placed as tasks in the Improvement Backlog, and the team will schedule their training in their iteration planning.

Establishing the Learning Parking lot enables a coach to prepare a coaching plan and creates an educational baseline. Then, as learning is completed, they can record this on their Evidence Board to remind them of their progress.

These activities are an initial assessment of the Team Context and one the team will finesse over time. This initial assessment enables the team to move into backlog preparation and then Mobilization and Planning as quickly as possible.

Forming a new team starts with the Context Assessment (see Figure 19). The required skills and team size depends upon analysing the Solution Goal. An assessment allows the assembly of an initial team which may be altered later as the activity progresses and new skills or the need for additional team members are

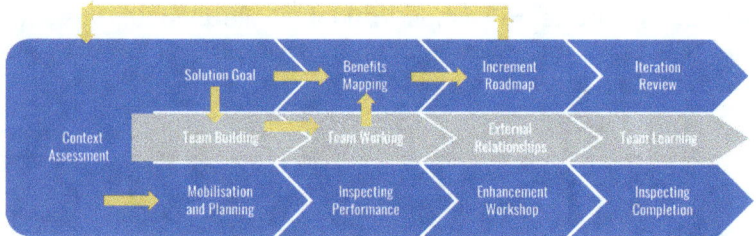

Figure 19 Forming a new team.

identified.

An evaluation of the organisational impacts of the Solution Goal will determine the involvement of external parties and stakeholders. The team size, skills and interaction with external parties will determine how the team should be organised and the type of Working Agreement required.

As the new team explores their context, they will quickly move into Mobilisation and Planning, where they will define how and when to create their solution.

When an agile team faces a new initiative, there are two likely scenarios; the required activity will either develop something new for the organization or enhance an existing solution, product or service. Agile Lineout has been designed to cater for both situations.

For new solutions, Agile Lineout starts with defining the Solution Goal. As previously stated, the Solution Goal influences the size and makeup of the team. Once assembled, a team will wish to explore the scope and prepare the backlog as quickly as possible.

In Benefits Mapping, the Solution Goal is broken down into features or outcomes and sized, probably using T-shirt techniques. In software activities, features are defined by acceptance criteria and broken down into user stories, non-functional requirements, and tasks. In non-software activities, the Objectives or outcomes are determined by measurable key results OKRs

The Solution Owner creates the Increment Roadmap, defining the sequence in which the solution will be delivered in viable increments. The Increments are stated in terms of dates and iterations. The time box horizons define the roadmap in strategic change initiatives, and tasks are identified and estimated. The resultant work items feed into the Mobilization and Planning activity.

The team will likely need to decide what to change when enhancing an existing solution (See Figure 20). In these circumstances, Agile Lineout starts with a Solution Review. This review is similar to an Iteration Review for a new solution build.

Strategic change activities should nearly always begin with a situational review to establish the organisational background before commencing the transformation activities.

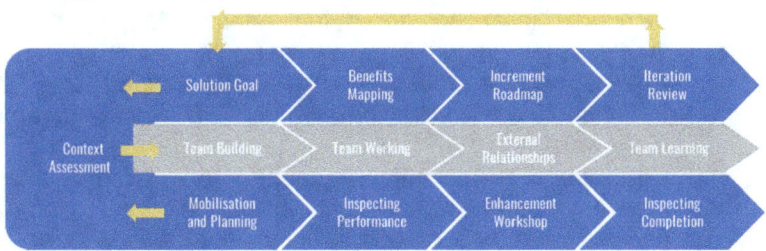

Figure 20 Enhancing and existing solution

The review doesn't need to be lengthy but sufficient to define the changes and new outcomes required from the enhancement initiative. In addition, this activity may clarify the Solution Goal and enable the team to scope and then sequence their initial actions.

Finally, the information from the review allows the team to proceed to backlog refinement.

A review is also needed for the Development Line when enhancing an existing solution. This review considers the tools, processes, and documentation used to deliver the previous Solution. The feedback from this review defines the starting point for the delivery practices for the enhancement initiative.

Sometimes, the agile ecosystem may have changed between completing the original solution and the planned enhancement. For example, new tools may have been introduced, new processes defined new regulations passed in the marketplace. All of these factors could influence the enhancement activity's working practices.

Coaching the Solution Line

In addition to Dr Deming's System of Profound Knowledge, the Solution Line in Agile Lineouts uses the Theory of Change from the social development sphere. The Theory of Change is termed a 'theory' because social development activities like business and cultural change are complicated and challenging to predict. The Theory of Change model has five elements:

- **Agree on what is needed.** This element is aligned with the definition of the Solution Goal
- **Challenge assumptions** This element is addressed in the Context Assessment and the Benefits Mapping activity.
- **Mitigate against risks** Risk identification, mitigation and monitoring are considered in Benefits Mapping and the creation of the Increment Roadmap.
- **Develop stakeholder ownership** Stakeholders create the Solution Goal and participate in the Iteration and Increment Reviews.
- **Enhance accountability** Accountability is a cross-over to the Team Behaviour elements in Lineout, where the team assigns roles and responsibilities during Mobilization and Planning.

In strategic change activities, a team can identify and analyse a range of interrelated elements by thinking through the Theory of Change to help ensure they are on the right path. A team can also challenge its fundamental assumptions and significantly mitigate risks. A well-planned Theory of Change helps ensure that the activity fits the purpose and will likely lead to the desired outcomes.

Tom Gilb talks of the consequences of ambiguity in the Solution Goal. I have found that a Solution Goal is not stated clearly enough on several occasions, so I have recommended defining the goals for strategic change as OKRs. To reduce the need for a team to seek clarification regarding the intent before setting out on the development journey.

In his PLanalysis book Tom Gilb suggests considering the following to get a much tighter definition of what is needed:
- **Identify ambiguous words**, terms or outcomes with unclear meanings.
- **Separate the link words**, such as and, but, or however. They suggest a relationship that requires an explanation and sometimes a source. Before and after the link word, both clauses require separation, exploration, and clarification.
- **Clarify** and, where possible, quantify generic terms such as people, productivity, and satisfaction.
- **Look for causality** where the subject of the Solution Goal may be the effect caused by an external or out-of-scope factor.
- **Look for side effects** and potential negative impacts of achieving the solution goal.

In the player's section, I outlined that the Solution Owner is a boundary role. I based this on materials suggested by Étienne Wenger in his boundary practice concept from his book Communities of Practice: Learning, meaning, and Identity (1998).

Wenger's boundary practice concept is based on asserting that an organisation could be viewed as a constellation of practices or functional silos. Boundary-spanning is the organisational practices, committees and the like that are joined together when developing a solution. The Solution Owner naturally amalgamates inputs from parts of an organisation to define the business requirements. Boundary processes are the objects which connect these silos. Boundary-spanning is an activity by which practitioners make connections, and boundary objects are processes or products that allow organisational practices to connect.

According to Etienne Wenger, the Solution Owner uses boundary practice to maintain connections between parts of an organisation. For example, they operate as liaisons between the customers and the team. In this role, they address conflicts, reconcile perspectives, and find resolutions.

The Solution Line also uses the theory of flow from Lean and links to the Development Line. In "Standing on the Shoulders of Giants", Eliyahu Goldratt, founding father of the Theory of Constraints, shows that improving flow should be the primary objective when reducing lead time. In recommending Kanban in Agile Lineout, I have provided the tools for a team to consider their flow. The Kanban may be used to evaluate work and wait for states so teams may have the data to assess their delivery efficiency. Flow theory breaks down the Solution into smaller incremental deliveries and creates the Increment Roadmap.

Coaching the Development Line

The Development Line uses all four elements of Deming's System of Profound Knowledge. In addition, it makes extensive use of the Deming Cycle. The Deming cycle is a continuous quality improvement model which consists of a logical sequence of four key steps: Plan, Do, Study, and Adjust.
- **Plan** is the identification of a path to achieve the solution or iteration goal.
- **Do** the work items selected from the backlog.
- **Study** considers the feedback from the Iteration Review
- **Adjust** the plan with actions necessary to keep the activity on its course.

The first step within the Deming cycle is understanding what you want to achieve and then planning how to get there. The planning is both a practical and theoretical step. In many instances, teams are not dealing with scientific or new-product discovery. Instead, they are dealing with a development process with a clearly defined outcome in strategic change activities. However, because of the human side of strategic change, while the backlog may be relatively stable, the incidence of unexpected activity with fires to fight is high.

The second step is to do whatever was planned. Next, Study the results to see if the desired outcome was achieved. Lastly, adjust the plan so the activities remain on course to achieve the goal. However, the study activity may suggest that the original intent was missed, is redundant, or offer another as an alternative.

The Deming cycle is a production control framework based upon rapid feedback loops, allowing a team to adjust its activities as it progresses to reach the desired outcome. The need for fast feedback determines the length of the iterations. The shorter the iteration duration, the better, but the iteration must also be long enough to produce something of value.

Mobilisation and planning

Mobilisation and planning is the team's development line strategy to decide their delivery process for the next Iteration. In mobilization, the team understands the solution context and evaluates their methods, tools and techniques for its creation, then sets out with the activities.

At the outset of an initiative, the team coach may need to make the team aware of the agile ecosystem provided by the organisation. Then, the team must decide which tools and processes are appropriate for their activity and develop their working practices accordingly.

However, too much focus on the ways of working in an early stage of the activity may not always be appropriate. Richard Hackman explains an experiment where they compared the performance of two sets of teams. The first group plunged straight into their task, but the second group was given a preliminary assignment of defining their working ways before starting. The results were that the plunge-right-in teams outperformed those which discussed their delivery strategy first. They concluded that the beginning of the activity was the wrong time for an intervention to look at ways of working or quality improvement.

Mobilisation and planning, therefore, should not be seen as a once-off activity but something considered at the beginning of each iteration. For example, as work progresses, a team may discover that alternative working practices in planning, prioritizing, sizing, and capacity management are needed. In addition, during Inspecting Performance, different ways of working changes may be revealed as constraints are identified.

Managing the Kanban Board

I apologise to those coaches who fully understand Kanban; however, I have found that while many coaches are well-trained in scrum, few are equally well-trained in Kanban. I have therefore outlined some of the basic principles here, starting with designing the Kanban Board itself.

The way to design the Kanban board is to define the high-level delivery system. For example, the system may be as simple as the following steps;
- define the task,
- break the work down so it can be completed in an iteration,
- do the work,
- verify the outcome,
- the work item is done or completed.

Next, construct two columns on the Kanban Board for each step except the last one. The first column is used when a work item is being processed. The second column is a holding area for the work item waiting to proceed to the next step.

Then, set work-in-progress (WIP) limits for each column apart from the final "done" step. WIP limits have two essential purposes in Kanban.
- Firstly, they limit the amount of work impacted by changing priorities, requirements or designs. This limitation can be of enormous value in cultural change initiatives when an activity may have unforeseen consequences that must be quickly addressed.
- Secondly, WIP limits restrict the flow and enforce the discipline of completing a work item before starting a new one. In addition, the finish-one-before-you-start-one logic increases the efficiency of the team.

It is, therefore, essential to make the items on the Kanban meaningful and trackable. For example, while providing "ongoing support" may absorb time, it is not trackable. Better to put the outcome of the support on the board, which may have a binary status of achieved or not achieved.

The two columns for each step on the Kanban board allow teams to monitor the ratio between work and wait-for states. Analysis of this type gives indications of capacity or process constraints, enabling a team to modify and improve its development process.

Using Kanban, it is needless for a team to discuss what they did, what they are doing and what they will do because it is evident on the board. This fact releases time in the daily Inspection activity for the team to consider their metrics and effectiveness and assess whether their plan and estimates will allow them to achieve the Iteration Goal. If the Team finds their goal will not be reached, their plans must be adjusted.

When using Kanban, Classes of Service are used for different types of activity. Classes of Service define rules by which the team will operate, for example, prioritising some types of work over others. As a result, the Class of Service may reduce the waiting times in the workflow for the high-priority items and makes their final delivery more predictable.

In the first Iteration, a team makes assumptions regarding its capacity, its use of estimation benchmarks or average-time-to-completion. Teams sometimes use past projects or metrics from other Teams undertaking similar tasks to create assumptions for the first Iteration.

After the first Iteration, the team has its own data to analyse its recent efficiency. It may then consider potential improvements in its delivery process. In this way, and as time passes, estimation and forecasts are, based upon the team's most recent delivery performance. Often a Team Coach will need to collect and summarise the data required using trends and graphs to facilitate the team's analysis.

Enhancement workshop

In *Leading Teams*, by J Richard Hackman, organizational psychologist Connie Gersick's findings explain why specific coaching interventions are helpful at different times in a team's task life cycle. Different coaching types of interventions with appropriate content are appropriate at each iteration's beginning, mid-point, and end.

Connie Gersick found that each team she tracked developed a distinctive approach to its task as soon as it commenced. That approach remained until almost halfway through the task duration. However, all teams transitioned at the mid-point, where they were willing to consider new perspectives to their work. As a result, they altered their ways of working to achieve their goal. The teams then adopted a finisher-completer focus until their task was complete.

An analogy Richard Hackman uses is a sports team that has just won a major trophy. Is this team in a mental state to reflect and learn? No, they want to celebrate! Similarly, a team that has just lost a significant trophy. Are they in a mental state to reflect and learn? Probably not because they are dealing with their disappointment and emotions of losing.

These are extreme examples, but we know that an agile team who has just completed a deliverable may be exhausted, have received negative feedback in the deliverable review, or want to get onto the next sprint. In these circumstances, are the team in a mental state for objective reflection and learning? Probably not! Could timing and team psychology be the key to ineffective retrospectives? Connie Gersick's research suggests that mid-sprint is better timing for a process performance improvement activity. I have therefore placed the Enhancement Workshop at the midpoint of the iteration to align with Connie Gersick's findings.

Coaching team behaviour

The essence of coaching team behaviour is that the individuals must trust each other.

Amy Edmundson defined the concepts of Psychological Safety in Psychological Safety and Learning Behaviour in Work Teams. Edmondson showed that team psychological safety is associated with learning behaviour. Interestingly she found that a team's learning behaviour mediates between team psychological safety and performance.

Peter Senge states, "*Team learning involves mastering dialogue and discussion, and these are the two distinct ways teams converse.*" In dialogue, there is the free and creative exploration of issues, deep listening to one another and suspending personal views. Different perspectives are presented and defended in the discussion, and there is a search for the best idea to support decisions or agreements. The team learns due to its ability to engage with each other positively.

These dialogues and discussions depend upon the team's maturity and the level of Psychological Safety that they have established. Amy Edmundson's analysis suggests that a coach needs to consider team structures, contextual support, team leader coaching, and shared beliefs to improve team performance and shape team outcomes.'

A team must understand that the intention of every interaction is developmental, with a positive outcome for all. The individuals must recognise that interventions are designed not to point blame or say what's expected of them but to expose and discuss their view of the reality of a situation.

The reality of teamwork often starts with establishing a team goal. The team goal is an agreement of what type of team the individuals want to be. For example, the team goal in soccer may be "We want to play Champions League Football"; in Baseball, "We want to win the world series". The team goal could be "we want to deliver a bug-free digital experience for our customers" in a business context. Setting a team goal sets a north star for a coach that gives a target and direction for team coaching.

Forming teams

With a new-to-agile organisation, resolving ambiguities about who is in the team and who is not may be necessary. For example, clarity is required regarding the people who previously had responsibility and accountability for outcomes in the old ways of working.

The introduction of agile may result in some duties from the traditional organisation transferring to the agile team. From a coach's perspective, care and clarity are often needed when dealing with individuals whose roles have been altered or displaced. These displaced individuals can either unintentionally or sadly intentionally derail an agile team's purpose, impact, and outcome. Therefore, a team coach must ensure that roles are clearly defined and step in to prevent processes from reverting to previous norms.

The importance of establishing team boundaries should not be underestimated. I have referred to Clayton Paul Alderfer's work. Alderfer is an American psychologist known for his Embedded Intergroup Relations Theory (EIRT). According to Alderfer. Relationships within a team have the following characteristics:

- Group boundaries,
- Power differences,
- Affective patterns,
- Cognitive formations, including distortions and,
- Leadership behaviour.

The boundaries around a group determine membership within that team and can be physical or psychological. A team must have clear boundaries to feel collective responsibility and emotional connectivity, an essential ingredient in agile.

Permeability is a concept that refers to how a team regulates and performs its transactions with other groups or parts of the organisation. Clayton Alderfer defines boundaries as permeable, describing how outside influences impact the team. Alderfer talks of three states overbounded, underbounded or optimally bounded. Interestingly, while groups without borders are complex to coach, overbounded teams may also be challenging from an organisational perspective.

An overbounded team may become elitist and have solid boundaries. It is obvious who is in the team and who is not with an overbounded team!

An overbounded team may assume they have the authority and remit to do everything themselves. The consequence of being overbounded is that a team may be reluctant to engage or use material support and capabilities from functions outside the team. Overboundness creates waste at both the team and organisational levels. Overboundness may be created if the team feels collectively in a very unsafe organisational environment.

An underbounded team, on the other hand, does not have clear boundaries and allows the outside organisational environment to impact its activities significantly. Picture a team being blown by the corporate wind! Such interference often happens when team members are not dedicated to the team's activity. Or in a newly formed team, members still have activities to complete from their previous roles. Being underbounded impacts team formation. Team members may feel that they do not belong or have conflicting priorities.

I have found that putting outside team commitments or tasks on the team task board creates transparency, reduces resentment for the time spent on non-team activities and increases the sense of one team.

Team building

Having established the team boundaries, a coach's attention turns to converting a group of people into a team. Team building is an often talked about yet poorly defined subject in agile circles. The team-building activity in Agile Lineout mirrors the team orientation and team leadership principles outlined by Dickinson and McIntyre.

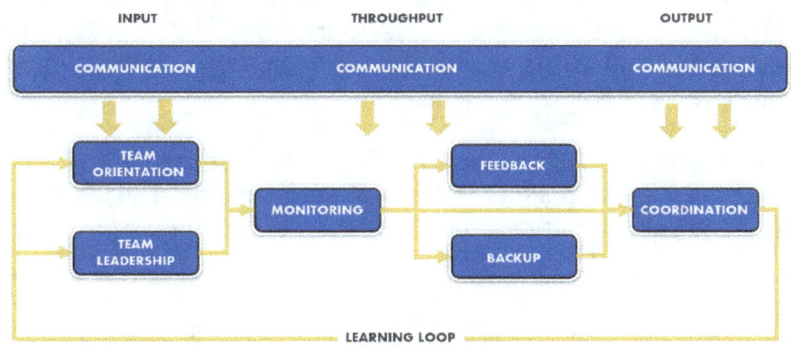

Teamwork model: Dickinson and McIntyre A Conceptual Model for Teamwork Measurement (1992)

Figure 21 A conceptual model for team work

The Conceptual Model for Teamwork Figure 21 shows team orientation and leadership and how team members develop attitudes and behaviours towards each other. A coach may use it to diagnose a poorly performing team.

Building upon Senge's principles, mentioned earlier, involving dialogue, Dickenson and McIntyre emphasise communication. Communication comprises exchanging information between team members differentiated at specific times in the activity journey. For example, they call communications at the beginning of the activity **Input,** during the activity **Throughput** and at completion **Output**. The learning loop should also be noted, as this shows that team building is a continuous process involving six measures.

- **Team orientation** - Refers to team members' attitudes toward one another.
- **Team leadership** - This does not refer to a manager role but how a team has informal leadership from within itself.
- **Monitoring -** Concerns being aware of their collective and individual performance.
- **Feedback** - Refers to the giving, seeking and receiving of information regarding performance and collaboration between team members.
- **Backup behaviour -** Involves assisting each other and breaking functional silos within the team
- **Coordination -** Refers to the team undertaking activities in a timely, coordinated and integrated manner

In the enhancement workshop, the team looks holistically at their performance and discusses ways to increase performance. This activity mirrors Dickinson and McIntyre's monitoring and feedback components of teamwork measures.

The Backup and Coordination components of teamwork measures happen during the daily inspection activity. The team may decide to pair, huddle or swarm to support each other and resolve a problem.

Team Behaviour also uses the Expert Coaching section of J. Richard Hackman's book *Leading Teams Setting the Stage for Great Performances*. Connie Gersick and Anita Woolley quoted in the Expert Coaching chapter and study of their original papers. Connie Gersick: *Time and Transition in Work Teams: Toward a New Model of Group Development* is used to decide when and what type of intervention should a team coach make. Anita Woolley: *Collective*

Intelligence: Three Factors for Building Smarter Teams focuses on how teams work together.

For further information regarding Agile Lineout, please contact Beneficial Consulting Ltd.

Jonathan Ward
Beneficial Consulting Ltd.
Dashwood House
69 Old Broad Street
London
EC2M 1QS

Web: beneficialconsulting.co.uk
Email: Jonathan.Ward@beneficialconsulting.co.uk

Mobile Tel: +44 (0) 7802 88459

www.ingramcontent.com/pod-product-compliance
Lightning Source LLC
Chambersburg PA
CBHW070422220526
45466CB00004B/1507